BRIDGES OF THE THAMES

RUTH & JONATHAN MINDELL

BRIDGES OVER THE THAMES

RUTH & JONATHAN MINDELL

BLANDFORD PRESS
POOLE · DORSET

First published in the U.K. 1985 by Blandford Press,
Link House, West Street, Poole, Dorset, BH15 1LL

Distributed in the United States by
Sterling Publishing Co., Inc.,
2 Park Avenue, New York, N.Y. 10016

British Library Cataloguing in Publication Data

Mindell, Ruth
Bridges over the Thames
1. Thames, River (England) – Bridges
I. Title   II. Mindell, Jonathon
624'.2'0942   TG57

ISBN 0 7137 1542 1

Typeset by August Filmsetting, Haydock, St Helens

Printed in Spain
by Grijelmo, Bilbao

CONTENTS

ACKNOWLEDGEMENTS

Besides the many books we consulted, which are given in the bibliography, we would like to acknowledge the following persons who were most helpful in the making of this book.

We would like to thank G. Watkins of Ealing Reference Library, and the staff of the libraries at the Guildhall, Reading and Twickenham. The various riverside innkeepers adjacent to bridges supplied us with many an ancient tale and relevant information. We thank David Wilson for answering our many questions between opening and shutting the lock gates at Godstow. We also thank Debbie Young for her assistance in a hot darkroom in Manchester and Anita Lawrence for drawing the maps.

The London Business School generously allowed us the use of its wordprocessing facilities.

And, finally, the person to whom we give our most grateful thanks – Bertram Mindell, for accompanying us on freezing November night photographic sessions, many helpful suggestions regarding the script, and the colour photograph of Richmond Bridge.

Ruth and Jonathan Mindell
London 1985

To Bertram

INTRODUCTION

Over the centuries many books have been written about the River Thames. Authors have told of the countryside through which the river flows, of the restaurants and inns which make welcome stopping points and of the interesting historical towns and villages along the Thames Valley. There have been books about walking holidays and boating trips, and about the history and geography of the Thames and its surrounding areas.

Many of the books have made mention of some of the bridges. They are the landmarks which make the Thames unique amongst the rivers of the world. Indeed there is no other river which has so many bridges crossing so few miles; although the Thames is only 142 miles long from Lechlade to Tower Bridge, there are more than 100 bridges between the two points. Some of them have been at crossing points for centuries, repaired or renewed from time to time. The reasons for their original construction are steeped in the history of Southern England and its ever-growing population.

Many different designs and materials have been used. They range from the simple single-arch structures near the source to the more complex feats of architecture required to cross the wide expanses in London.

Before the thirteenth century bridges over the Thames were made of wood. Huge rough piers and piles were fixed into the river bed and the road was laid out above them. These wooden bridges were often destroyed or severely damaged by the winter floods and the fierce scour of the water. Eventually it was decided that bridges should be made of more permanent materials and in 1209 London Bridge was opened – built of stone.

Ha'penny Bridge and the Toll House at Lechlade.

Six centuries later, as a result of the Industrial Revolution, iron and subsequently steel were used in bridge construction, particularly for railway bridges. In recent years pre-stressed concrete has been used to withstand heavy motorway traffic.

The earliest bridges were built for strategic reasons – to provide crossings for armies. Although some of them were destroyed during battles (such as those damaged in The Civil War) or collapsed due to neglect, they were all rebuilt. Later, when the land was at peace, bridges were built to aid the flow of goods and commerce and help the needs of a growing population.

During the last 150 years the major reason for bridge building has been the increased movement of people crossing by horse, railway and, this century, by motor car. Even today there are plans for one more bridge to alleviate London's ever-growing traffic congestion. There are architectural plans for a new bridge at Thamesmead – while, by contrast 150 miles up-river, the old bridge at Radcot stands after more than 700 years . Where else does one have such a kaleidoscope of history? Many bridges over European rivers were destroyed during World War 2. By contrast, the only Thames bridge to suffer war damage was Waterloo.

One more question must be answered before the history of each bridge can be explored. Who controls the Thames? Who is responsible for its navigation, its water systems, its embankments – in other words – its welfare? Formerly the Thames from source

Under the navigation arch of St John's Bridge looking towards Lechlade and the Church of St Lawrence.

to sea was controlled by many different authorities. This did not always make for either efficiency or co-operation. The City Corporation, which had looked after the river for 600 years, having for many of them obstructed the development of bridges, relinquished its powers in 1857. In that year its authority was transferred to the Thames Conservancy. In 1908 an Act was passed creating the Port of London Authority which became responsible for the Thames from Teddington to the sea. In April 1974 control of the river from the source to Teddington was transferred from the Thames Conservancy to the newly formed Thames Water Authority – including a Port of London Authority Division as well. Thus today the Thames Water Authority controls and administers the whole river from source to sea.

Our journey commences at Lechlade, the navigable head of the river. Through both narrative and photographs we take the reader downstream towards London. In each chapter the descriptions of the bridges are preceded by an introduction to the general history and geography of the particular section through which the river flows.

1 FROM SOURCE TO OXFORD

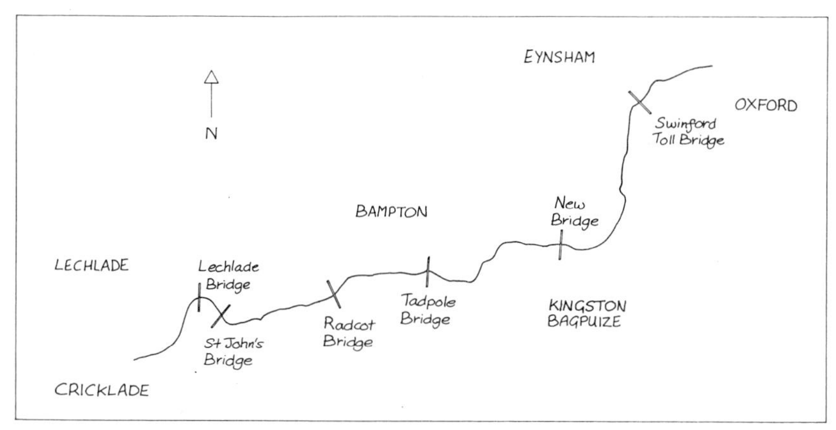

The Thames from Lechlade to Oxford. (27 miles).

Archaelogical finds in the Upper Thames region show that as far back in history as the early Stone Age man lived in settlements along the banks of the River Thames and its tributaries.

The roots of present day thriving Thames-side communities like Lechlade, Oxford, Abingdon and Wallingford can be traced back to the time when the Thames Valley was joined to the continent of Europe. In those days the Thames was merely a tributary of the Rhine.

How did primitive man cross the river? There were no bridges in those distant times. How did he travel up and down the river? The land near the river was often marshy or so thick with vegetation that any land travel would be difficult. It seems likely that one of man's earliest observations must have been that wood floats. If he were to sit astride a large log he could propel himself

The weir at St John's Bridge.

across, or indeed up and down the river. It would not have taken many more years to discover that if he tied a few logs together he would have a simple raft to take him over the water.

In the winter when the river is full and flows fast, it would be hazardous. But in the summer when the river is low and the current light it would be easy to cross the river in this way. In the summmer there were also several parts of the river where natural fords existed. When the river was low, these shallows provided crossing places.

Duxford was such a ford. In Roman and earlier times Duxford must have been an important place because it was one of the few localities where the river could be crossed in this way. Today, Duxford is a tiny village situated about two miles above where the Windrush flows into the Thames at Newbridge. Only a track leads down to the river.

Shillingford and Wallingford (below Oxford) were more important crossings than Duxford. They were situated on old trade routes and, as agriculture developed, they expanded in Roman and later in Saxon times.

These natural crossings were not only important for trade routes but they were strategically vital to the warring factions from Roman times until those of Cromwell.

Over the centuries the River Thames provided transport to markets by the links of its fords and later its bridges with the roads that developed from pre-historic trackways. Food was the main source of wealth: grain, cheese and meat. The Vale of the White Horse was famous for its cheeses as long ago as the Domesday book in 1086 . Most of the Vale's cheeses were carted to Buscot and then shipped to Oxford and London. In 1809 it was noted that nearly 3,000 tons of cheese was sent from the 'Cheese Wharf' to London via the Thames.

The sheep farming that was introduced to the Oxfordshire Cotswolds by Neolithic man was developed throughout Roman and Saxon times. By the twelfth century large quantities of wool were produced which became England's main export during the Middle Ages.

Stone quarrying was another industry. The soft grey-yellow colour of the stone quarried around Burford and Taynton was used not only for Cotswold villages, but also for some of the loveliest bridges over the Thames.

The medieval towns such as Cricklade, Oxford and Wallingford grew to be such vital centres of commercial activity that in the late-Saxon times they were granted a royal licence to mint coin.

As the towns by the Thames grew in population so did the need for better communications. The roads could not stop short at the river. Somehow, the water had to be spanned. Traffic could no longer depend on the vagaries of the seasons. The river had to have permanent safe crossings, even when it was in flood with swift swirling currents. Safer methods had to be invented and so at last we came to the bridges that have been designed and built over the centuries up to the present day.

The source of the Thames has been disputed since earliest times. Some say that its life begins at Seven Springs and that the River Churn, which flows from there is really the beginning of the Thames. Others maintain, and this is the majority view, that the source bubbles up in a field very close to Cirencester – appropriately called Thames Head.

Having risen in Gloucestershire it is no more than a stream until it reaches Cricklade in Wiltshire. In the days of Alfred the Great, the ford at Cricklade was one of his strategic boundaries. Today, though, the Cirencester to Swindon road no longer depends on its ford because the river is spanned by a single-arch steel and

The span over the new navigation channel at Radcot.

concrete bridge. In contrast, a few yards away, there is a footbridge made from a telegraph pole that leads to a cottage.

Although there are a score of similar small footbridges leading to individual cottages above Waterhay and Cricklade Town Bridge, it is not until Lechlade is reached that the bridges and the history of their development became of interest. Lechlade has been the effective Head of Navigation for a long period. Above Lechlade the river is not easily navigable by anything larger than a canoe, although small motorised craft can, with care, go a mile or so upriver as far as Inglesham Tower, after which the water becomes extremely shallow.

So, this first section takes us from the Head of Navigation at Lechlade to the outskirts of Oxford, through some of the most peaceful of Thames-side country.

LECHLADE BRIDGE

After the Thames and Severn Canal was opened in 1789, Lechlade became a very busy township. Apart fom the ancient ferry which pedestrians used, all other traffic wishing to cross the Thames had to go further down river to St John's Bridge. This caused long delays and much inconvenience. So the people of Lechlade decided to petition Parliament to authorise the building of a bridge to replace the ferry.

The first petition failed, but in 1792 Parliament passed an Act authorising the building of a bridge and also the improvement of the town's road systems. Burford and Swindon districts each appointed a set of trustees who together would be responsible for building and maintaining the bridge. The bridge was formally known as 'Lechlade Town Bridge', but locally it was nicknamed 'Ha'penny Bridge'. This was because when the bridge was

Lechlade Bridge (known as Ha'penny bridge).

completed tolls were exacted from the users on the following scales. Pedestrians were charged a halfpenny unless they were on the way to church or were mourners! Carriages were charged 6d. and a horse 2d.

From early in the nineteenth century petitions were collected by Lechlade townsfolk in protest against the halfpenny toll. However, it was not until the Act of 1853 that Parliament abolished the pedestrian toll and the toll on horses. In future only 'carriages driven by steam or other machinery' were to be charged at the rate of one shilling per wheel.

On 1 November 1875 the toll charges were abolished entirely. The local authorities became responsible for the bridge maintenance in place of the trustees.

The bridge itself was built from locally quarried stone. One half is in Wiltshire, the other half in Gloucester. The navigation arch spans 40 ft and was specially constructed to allow barge traffic to pass unimpeded. The architect, Daniel Harris, also designed a narrow towpath arch for the men on horses pulling the barges, and a low floodwater gate on each side of the bridge. Although no tolls are charged today, the toll-keeper's cottage is still attached to the bridge on the Gloucestershire side.

Legend has it that when the master mason completed the bridge he climbed up a tall tree by the Inglesham Round Tower saying that if the bridge collapsed he would throw himself into the river! Fortunately the bridge withstood the weight of the traffic and has done so ever since.

In 1973 the Gloucestershire County Surveyor ordered the bridge to be renovated as weaknesses in the stone had become evident. £3,500 was spent on renewing and strengthening the foundations so that today even the heaviest of lorries may pass over it.

ST JOHN'S BRIDGE, LECHLADE

With the exception of London Bridge, St John's Bridge was the earliest stone bridge over the Thames. The original single-arch bridge was built by the monks of St John's Priory in 1229, replacing the pre-existing ford.

St John's Lock Cut Bridge viewed from the lock.

In 1234 St John's Bridge Fair was established by a charter. It was held in the field adjoining the bridge and people came from beyond Oxford to buy and sell their wares.

Thomas Baskerville, in 1692, wrote an account of 'Bridges over ye famous River Tems from Cricklade to Wallingford'. He said that 'St John's Fair is kept in the meadow below ye bridge on Gloucestershire side to which Oxford boats and others resort to sell Ale, Beef and Carrots... More especially sage cheeses in various shapes and colours, which I have scarce seen anywhere else'.

In 1341 pontage (the right to levy tolls) was granted to the incumbent prior for maintenance and repair to his bridge. Further pontage was granted in 1387 when the bridge needed major rebuilding after it had been destroyed by an order of the King's uncle, the Duke of Gloucester.

In 1677 the people of Cricklade, to which barges went on from Lechlade, complained of the piratical activities of a certain Captain Cutler who held barges to heavy ransom at St John's Bridge!

In 1751 an Abingdon order declared that 'every boat passing St John's bridge pay for every five tons the sum of 3d'. In 1790 a pound lock was built at St John's because of the additional traffic originating from the Thames and Severn Canal. A bridge had to be built over the resulting lock cut – but it was so poorly constructed that by 1795 it was in a state of collapse. It was repaired several times during the following century, but by 1878 it had to be completely restored.

The upper part of the present bridge is little more than 150 years old, having been extensively repaired in 1820 after 500 years of service.

It is thought that the present day Trout Inn (called St John the Baptist's Head for centuries) may have been built with some of the stone from the original Priory of St John the Baptist.

RADCOT BRIDGE

After Lechlade the river winds through unpeopled countryside. The road over Radcot Bridge (built over an ancient trackway) joins Bampton in the north with Faringdon to the south. Both market towns lie a couple of miles inland from the river.

According to the Victoria County History this picturesque bridge may be more than 700 years old. Part of the masonry is

Radcot Bridge – the oldest surviving brickwork over the Thames.

believed to be the oldest original stone work left across the Thames. In records of 1312 a grant of pontage was allowed for the repair of the bridge. This suggests that it must have been built some years before that date.

The bridge is about 22 yards long with a steep rise to the crown. It has three pointed Gothic arches and cutwaters on both sides of the bridge. Cutwaters were often added to medieval bridges to prevent the stonework being eroded by the movement of the water. In the centre of the downstream parapet there is a pedestal which is believed to have held a stone cross or the figure of a Virgin.

In 1387 one of the arches was destroyed by John of Gaunt's son, Henry of Lancaster (later Henry VI), during his revolt against Richard II. The King's men, under the leadership of the Earl of Oxford, arrived to find that it had been destroyed. Not only was the bridge impassable but Henry had a force of 5,000 men to support the cause. In the battle which followed the King's men were beaten, many losing their lives in the river.

During the Civil War Radcot Bridge was again the scene of several battles. Charles I's headquarters were in Oxford, not far from Radcot, therefore the bridge was a vital strategic point for the armies. In 1645 the heavily guarded bridge was attacked twice by Cromwell without success. On the third occasion it fell to him and Parliament.

In 1787 a second bridge was built at Radcot. It is a single-arch structure and spans the 'new' navigation channel. It was built to facilitate the increased water traffic. In reality, though, it is more difficult to pass clearly through this arch than the centre arch of the old bridge, because the new arch was built on a skew to the flow of the river.

Building stone from the Taynton Quarries near Burford used to be carted by road to Radcot and shipped from there. It is recorded that stone for parts of St Paul's Cathedral was brought to Radcot, put on rafts and floated down the river to London.

Berkshire cheeses were shipped from here as well as from Buscot. Until the coming of the railways, coal was also shipped from Radcot.

TADPOLE BRIDGE

As to why this bridge was built or by whom is not recorded anywhere. Nor do we know exactly when it was first constructed. Originally it may have been made of wood, but in 1802 the present stone bridge was built next to Tadpole Weir. It is a single-arch structure decorated by two roundels on each side. There is an 1894 floodmark on the bridge, which is well above the surrounding fields.

The bridge lies three miles downstream from Radcot, and carries the road from Bampton in the north to the Oxford to Swindon trunk road in the south. As coal was distributed from wharves next to Tadpole Bridge, it may be that the bridge was built to alleviate the traffic at Radcot.

There are no houses nearby; the only building is the small inn of which Fred Thacker in *The Thames Highway* wrote, '... at one time over the entrance was the legend: The Trout kept by A. Herring'.

NEW BRIDGE

It marks the confluence of the River Windrush with the Thames on the north side of the bridge and links Witney in Oxfordshire

New Bridge and the Maybush Inn.

Reflections at Tadpole Bridge.

with Kingston Bagpuize to the south. At each end of the bridge there is an ancient inn – The Maybush on the southside and The Rose Revived on the north bank.

Until about 1250 the Thames at this point was crossed by a wooden bridge. Then stones were brought down from Taynton quarries and used in the construction of this beautiful six-arched Gothic bridge. Along the upstream side there are triangular cutwaters, or buttresses, which reach up to the road level to protect pedestrians from the road traffic.

The earliest written record of New Bridge was in 1279. A list of employees at Standlake Manor nearby mentiones 'Thomas at Pontum'. meaning 'Thomas at the Bridge', probably referring to the toll keeper. The tolls were levied until 1600. Repairs to the bridge were recorded in 1415.

In 1644-45 several battles took place at the bridge, between the Roundheads and the Cavaliers. The King's men finally lost the bridge when they ran out of powder and shot, the bridge itself being partially destroyed.

The centre arch of Swinford Toll Bridge.

Since that time it has been so well maintained that in 1981, when Oxfordshire County Council surveyed the bridge, they concluded that it was fit to take even 32-ton lorries and that it was good for the next 300 years!

SWINFORD BRIDGE

The ford at Eynsham was often a perilous crossing because of the changing currents. In the Middle Ages the Benedictine abbots of Abingdon established a ferry between Eynsham and Swinford. In the winter months, though, the ferry proved to be as dangerous as the ford.

It is recorded by Thomas Crosfield of Queen's College, Oxford that in the winter of 1636 a party of Welsh sheriffs bringing ship money to Charles I crossed by ferry. The currents were so treacherous that the ferry overturned. Although eight men managed to swim to safety, four men drowned and £800 was lost.

In 1764 the preacher John Wesley had a narrow escape when he and his horse barely reached the safety of the bank, such was the force of the water.

With a view to building a toll bridge the fourth Earl of Abingdon bought land on either side of the ferry. His friend and financial adviser, Sir William Blackstone, MP for Hinton, Wiltshire, petitioned for the building of a bridge to replace the ferry. The Act of Parliament authorising the building was passed in 1767 and the bridge was opened to traffic just two years later.

The earl and his descendants were granted the right to levy the same toll as before for the ferry. In addition, vehicles were charged at the rate of 1d. per wheel. The Act permitted the earl to lower the tolls but not to increase them! In celebration of his son's marriage in 1853 he abolished the toll for pedestrians. In 1900 the toll for a bicycle was reduced from 2d. to a halfpenny.

In 1981 Mr Michael Cox and his wife Stella bought the bridge for £100,000. He applied to the Secretary of State for Transport to raise the toll from 2p to 10p. He claimed that his takings did not leave him sufficient profit to pay for the maintenance bills of £350,000. Permission was refused, however, and today the toll for motor vehicles remains at 2p.

It has never been definitely established who built the bridge, but there are indications that Sir Robert Taylor, a friend of Sir William Blackstone was responsible for the design and construction. Sir Robert was the son of a London stone mason who studied architecture in Italy. The Italian style balustrades and graceful symmetry make it likely that Sir Robert was the architect of this lovely bridge. Altogether it has nine stone arches. The central navigation arch is 36 ft wide. The arches on either side of the central arch are 27 ft wide. The six floodwater arches decrease in width from 24 ft to 20 ft and finally to 16 ft.

2 OXFORD

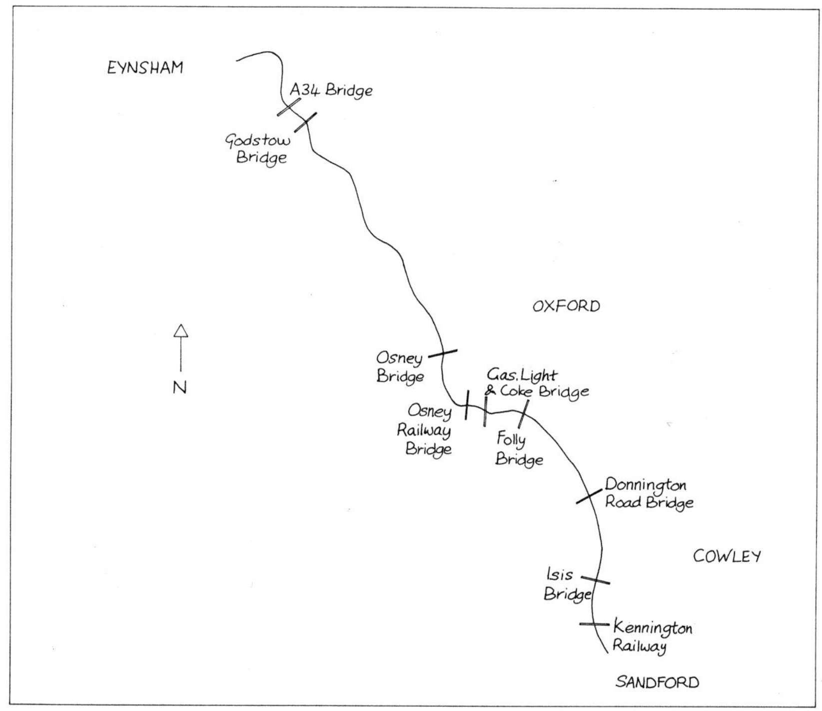

The Thames from Eynsham to Sandford. (9 miles).

Why do nine bridges cross the Thames around Oxford? It is not a very long stretch of river to be spanned at such frequent intervals. To discover the reasons one must go back hundreds of years and then travel down through the centuries to modern times when the most recent bridges were built. There are three 'old' bridges, four 'modern' bridges and two railway bridges.

Abingdon, Dorchester and Wallingford were all towns of note before the infant Oxford appeared as a dot on the map. They lay on the pre-historic trackways of the chalk downs and led to natural crossings at Wallingford and Streatley. The town of Abingdon was conveniently situated at the confluence of the River Ock with the Thames. As far back as neolithic times man had settled and traded there. It became a centre of Saxon Christianity when Cissa, a Saxon king founded Abingdon Abbey in AD 675. At that time, and for many centuries thereafter, all learning was rooted in the Church. Only those connected with religious foundations were literate. Therefore, Abingdon was a seat of learning long before the earliest recorded religious foundations existed in Oxford. St Friedeswide, an early English saint, who died about AD 735, was said to have founded the priory at Christchurch.

Oxford's situation, on the Thames and the Cherwell gave it strong physical defences. Its position also opened out important trade routes westwards to the Cotswolds and South Wales, northwards to Banbury and the Midlands and South East to London. Therefore, whoever held Oxford controlled these routes.

From the tenth century onwards Oxford grew rapidly. Alfred the Great's grandson Aethelston, who reigned from 924 to 939,

The former navigation channel at Godstow, now leading to the weir.

Godstow Bridge and weir from the terrace of the Trout Inn.

had coins minted there. Oxford soon became the most important town in the region. When unity became established in England, superceding the petty kingdoms of the Saxons, the Thames developed as a highway. It was no longer merely a barrier between warring factions.

Because of Oxford's growing importance routes to and from the city had to be improved. In 1004 a bridge was built across the River Cherwell. The ford at Hinksey was supplemented by ferries. The first historical record of a Thames bridge at Oxford was 1085 when Robert D'Oyley (one of William the Conqueror's principal Knights) built the bridge where Folly Bridge stands today.

A nunnery was built at Godstow in 1138 by Editha, widow of Sir Peter Lamelyne. Close by the ruins of the nunnery stands Godstow Bridge.

In the early thirteenth century the religious foundations in Oxford flourished, and the great orders of the Dominican and Franciscan friars who taught there attracted many students. The degree for which the men studied was an ecclesiastical licence to teach. The three subjects taught were Latin grammar, logic and philosophy. By the middle of the thirteenth century the student body had grown to 1,500 men living and studying in Oxford, which was a large number for a medieval town.

The fourteenth and fifteenth centuries saw the development and broadening of more colleges and halls of residence for the students. As more people came to live in Oxford so trade increased. The volume of goods carried along the river increased too. Even after the coming of the railways, the barges still carried heavy goods and all varieties of produce up and down the Thames.

Until this century Oxford remained a university town positioned at the centre of important trade routes, carrying market produce to all points of the compass in this mainly agricultural region. But in the twentieth century the beauty of Oxford was to some extent marred by the development of the motor-vehicle industry. W.R. Morris (subsequently Lord Nuffield) happened to live in Oxford and, because of this thread of chance, the city has developed into the urban sprawl which it is today. It is precisely because of this tremendous growth of population and its numerous functions that Oxford requires its many bridges to lessen the congestion which would otherwise bring its traffic to a halt.

A34 OXFORD BY-PASS AT GODSTOW

This is the first of the modern pre-stressed and reinforced concrete bridges to span the upper Thames. Its dual carriageways help to alleviate the traffic problems of the confines of the city centre. The

A34 Bridge at Godstow.

bridge was designed by R. Travers Morgan & Partners and constructed by Higgs & Hill, Ltd at a cost of £110,000. Opened in 1961, it not only forms part of the Ring Road around Oxford but also follows the route (as the A34) of one of Britain's oldest cross country routes from Manchester to Winchester.

GODSTOW BRIDGE

The Trout Inn and the ruins of the nunnery form part of the most romantic legend connected with any bridge over the Thames. The nunnery was built in 1138 by Lady Editha in memory of her husband Sir Peter Lameleyne. Lady Editha had a vision telling her to go to the meadow by the Thames and build a holy place for gentlewomen. The consecration of the nunnery was attended by King Stephen, his bishops and courtiers. It was rumoured that the nunnery was not so much a holy place for nuns, as a place of education for high born young ladies!

Among their number was the beautiful Rosamund, daughter of Lord de Clifford. Henry II chanced to meet Rosamund when he visited the nunnery. He fell in love with her and took her as his mistress. To meet Rosamund privately it was said that the King built a secret labyrinth which led to a bower. Legend has it that Queen Eleanor, learning of this liaison, copied the Greek tale of Theseus and the Minotaur. She tied a silken thread to Rosamund's dress. As Rosamund walked along the labyrinth the silken thread unwound and showed Queen Eleanor the way to the secret meeting place. The Queen confronted Rosamund and offered her death by dagger or poison!

Romantic as the story is, it was not altogether accurate. It seems more likely that Rosamund, on being discovered, retired to the nunnery. According to old records Rosamund appears to have died of natural causes some years later. She was buried in a magnificent tomb, which vanished at the time of Henry VIII's dissolution of the monasteries.

The date at which the old bridge by the Trout Inn was built is not known. However, one of the two arches is pointed. It closely resembles the old arches of Radcot Bridge. It is believed that it must have been built not long after Radcot as Godstow was a

Christmas Day at Godstow Bridge.

ERRATUM
Please note that the colour photos on pages 31 and 50 have been transposed.

frequently used crossing to the west of Oxford. The widest arch in the old navigation channel was twelve feet. As barges were eleven feet nine inches wide, it did not leave much room for error!

The new bridge at Godstow was built in 1788 after a survey by the Thames Commissioners. Although it was designed with two navigation arches, they were reported as being 'too low for laden barges to pass through'. Until 1892 the new bridge was kept in repair by the Earls of Abingdon, since when it has been maintained by the Oxfordshire County Council.

OSNEY BRIDGE

Originally Osney Bridge, or Botley Bridge as it was first called, was built of timber. After 1129 when Osney Abbey was founded it was thought that the monks built a bridge of three stone arches.

Osney Bridge.

The exact date when the bridge was built is unknown, but it stood until 1767. As the approach to this bridge was often flooded, a causeway was built in the early sixteenth century and was maintained by the Abbot of Osney. After the dissolution of the monasteries the causeway became so neglected that few people dared to cross into Oxford over the bridge for fear of drowning.

Sir William Blackstone promoted the building of both Osney and Swinford bridges and the repair of the roadway. He had many setbacks in obtaining the finance that was required for the two bridges and the restoration of the Botley Causeway. Finally, Sir William had to contribute £2,200 himself so that the road could be ready when Swinford Bridge was opened to the public. Until 31 December 1868 Osney was a toll bridge. On that day the ever unpopular tolls were abolished.

According to Fred Thacker, in *The Thames Highway*, 'In November 1869 the Thames Conservancy was seeking powers to rebuild the bridge, as it was a serious obstruction in its damaged state'. Thacker continues by saying that the bridge was in such a condition because 'Squire Campbell of Buscot had three barges wedged into it, and Charlie Bossom had to dislodge a boatload of stone from the bridge before they (the barges) could be released.' This naturally made the bridge most unsafe.

Nothing was done about the dangerous condition of the bridge

The coat of arms of the Oxford Local Board which authorised the building of Osney Bridge.

and then, in December 1885, tragedy struck. One morning a large part of the central arch of the bridge collapsed. A man and two small girls were catapulted into the bitterly cold water. The man and one small child were saved, but the second girl, Rhoda Miles, was drowned.

A temporary wooden bridge was erected some way downstream, but the necessity of a reconstructed bridge in the old position became apparent because of the numbers of people who still used the old broken bridge, and the accidents and disputes it caused.

Work on the new bridge began early in 1888. Mr M. White, an engineer of the Oxford Local Board, responsible for the construction, designed a bridge with a 60 ft iron span and two stone arches each of 30 ft. Because of difficulties with the piling the date for completion was postponed repeatedly. To add to the general feeling of malaise over the building of this bridge, on 18 November 1888 Rhoda Miles' body was found close to the weir – some three years after she was drowned. The new Osney Bridge was finally opened to all traffic on 31 December 1888.

In some ways Osney represents the separation point between the Upper Thames and the rest of the river to London. Above Osney there is a gentle rural atmosphere. One meanders slowly along the river between the field and flowers. Below Osney there are the riverside towns, the regattas, the large paddle steamers and the hordes of tourists. The reason for this separation is the single iron span of the bridge. It has been constructed so low that only small craft can pass under it and go on to the upper reaches. As a result the majority of craft do not venture further than Folly Bridge, on the downstream side of Osney Lock.

OSNEY RAILWAY BRIDGE

This Thames crossing provided the final link for the Great Western Railway's (GWR) 9½-mile branch from Didcot to Oxford which opened on 12 May 1844. That the opening came so long after the initial proposals of 1833 was principally the result of the views of the Oxford City Fathers and the college fellows who felt that the railway would be a damaging influence. As was typical of

Osney Railway Bridge.

hostility to the early expansion of the railways, the University felt that the railway would corrupt students' morals. The Oxford Canal Company organised opposition for somewhat more commercial reasons!

Today rail travellers from London cross the Thames for the sixth time at Osney Bridge. Of these six, Osney is of the simplest design when compared to I.K. Brunel's other masterpieces downstream.

There are in fact two bridges on this site. Passenger trains use the original crossing. In 1887 a second span was built immediately downstream to accommodate tracks that eventually led to the gasworks. When these lines were closed, the tracks on this newer bridge were converted to become goods lines, allowing passenger trains better access into Oxford station, just a few yards north of the bridge.

In 1983 much of the decking on the main span was renewed. At the same time the bridge received a new coat of green paint, which, while not as dazzling as the colour scheme of the Gas, Light & Coke bridge, is certainly a great improvement on the rusting and peeling grey that it replaced.

The newly painted Gas, Light and Coke Bridge.

THE GAS LIGHT & COKE BRIDGE

This unique crossing was built in 1886 to serve three purposes. In addition to carrying five 18 inch gas mains across the Thames, this 22 ft span allowed space for a single track railway and a narrow roadway into the gasworks themselves.

After the gasworks closed, the branch line and the gas mains were removed. Today, the bridge's sole remaining purpose is as a pedestrian link into the city centre.

In 1981 its wrought-iron girders and supporting cast-iron cylinders were repainted. The bridge forms an attractive distraction from what is otherwise one of the ugliest stretches of the upper Thames.

FOLLY BRIDGE

Over the centuries the name of this bridge has changed more than any other on the Thames. As far back as Saxon times a timber bridge existed where Folly Bridge now stands. But bridges made

Folly Bridge from the Head of the River Public house.

of wood needed constant repair and, as this was such an important crossing, a more permanent structure was required.

In 1085 Robert D'Oyley, with the assistance of masons from Abingdon Abbey, replaced the timber bridge with one of stone. In Norman times it was called 'South Bridge' or 'Grand Pont' as it was thought to be 'great' by the standards of those times.

Attached to the southern end the monks built a small chapel to St Nicholas. The money collected at this chapel was used for the maintenance of the bridge. The tower which became the 'Folly' was said to have been built in 1142 by King Stephen as a pharos or high watch tower for the defence of Oxford against his cousin Queen Matilda.

In the thirteenth century a Franciscan friar named Roger Bacon added to King Stephen's tower and converted it into an observatory. Here he spent nights noting the stars and writing on philosophy. He had a great reputation as one of the most learned men in medieval times. There were those who charged him with other special powers. Mr and Mrs Hall in their book of the Thames, first published in 1859, wrote, 'Popular prejudice accused him of practising magic and he was called to Rome by the General of his Order, but having cleared himself he was sent back to England'. At that time the bridge was called 'Friar Bacon's Study' or simply Friar's Bridge.

Another legend concerning the Friar attached itself to the tower. It was said that when a man of greater learning than Bacon passed under it, the tower would collapse! Although it decayed it never fell, but was eventually pulled down in 1779 by the city corporation as a dangerous obstruction to the road.

After Bacon's lifetime the tower was leased to a man called Welcome. Mr Welcome added another storey to the top of the tower, from which the bridge earned the new name of Welcome's Folly. When the tower was dismantled it was sold for a mere £13. The remainder of the bridge stayed as it was until 1825. Between 1825 and 1827 it was rebuilt in its present form. Although Mr Welcome has long been forgotten, his 'folly' gave the bridge its name up to the present date. People who are unaware of Mr Welcome's existence think the name a satire on the pleasure seeking life of the university students to whom boating is more important than their books!

When the bridge was rebuilt between 1825 and 1827, Ebenezer

A Thames Conservancy Launch passing under the backwater arch of Folly Bridge.

Parry designed it. There were 30 ft spans over the main river, a 21 ft arch over the navigation channels and smaller floodwater arches. The tolls were charged at the Hinksey turnpike gate about half-a-mile from the start of the causeway.

When the Great Western Railway opened in 1844 and the station was built nearby it was decided to build a toll house by the bridge itself. The increased traffic from the railway could then be charged as they crossed over the bridge. In August 1850 the tolls for crossing Folly Bridge were finally abolished, but the little toll house still stands on the northern end of the bridge.

Two landmarks adjoin the bridge. On the city side stands 'The Head of the River', a well-known public house. On the other side is Salters, one of the oldest and best known pleasure craft operators.

DONNINGTON ROAD BRIDGE

This bridge replaced a once well-known landmark, the Free Ferry Footbridge. The concrete bridge was designed as early as 1955, but was not completed until 1962. The consultant engineers were R. Travers Morgan & Partners and construction was carried out by The Cementation Company. It was opened on 22 October 1962 when Viscount Hailsham unveiled a commemorative plaque which is still in place in the parapet of the bridge.

Donnington Road Bridge.

ISIS BRIDGE

This rather dull single-span bridge forms part of the ring road of dual carriageway around Oxford. Like its sister bridge on the northern side of the city, near Godstow, it was designed by R. Travers Morgan & Partners who seem to have somewhat of a local monopoly in bridge designing. The name is taken from Isis lock just upstream of the bridge. The name 'Isis' is itself an alternative name for the Thames in the region of Oxford.

Isis Bridge.

KENNINGTON RAILWAY BRIDGE

The present girder bridge was opened for use on 29 July 1923. It was constructed of six 83 ft long steel bowstring girders, set in three arches resting on two pairs of cast-iron cylinders sunk into the river bed. The abutments of the original iron bridge, opened in 1864, can still be seen on the upstream side of the existing bridge.

When the bridge was opened it carried the GWR line from Princes Risborough through Thame to Oxford. Interestingly enough, this branch, with its continuation on the main line from Princes Risborough to Paddington, formed the shortest rail route between Oxford and London. Designed as a branch line, though, it did not offer the fastest services and did not see many through trains.

Since Beeching's axe in the early 1960s, this line has been closed in its middle section. Today, the branch continues for less than one mile away from the main line at Kennington Junction to British Leyland's plant at Cowley.

Kennington Railway Bridge.

3 ABINGDON TO READING

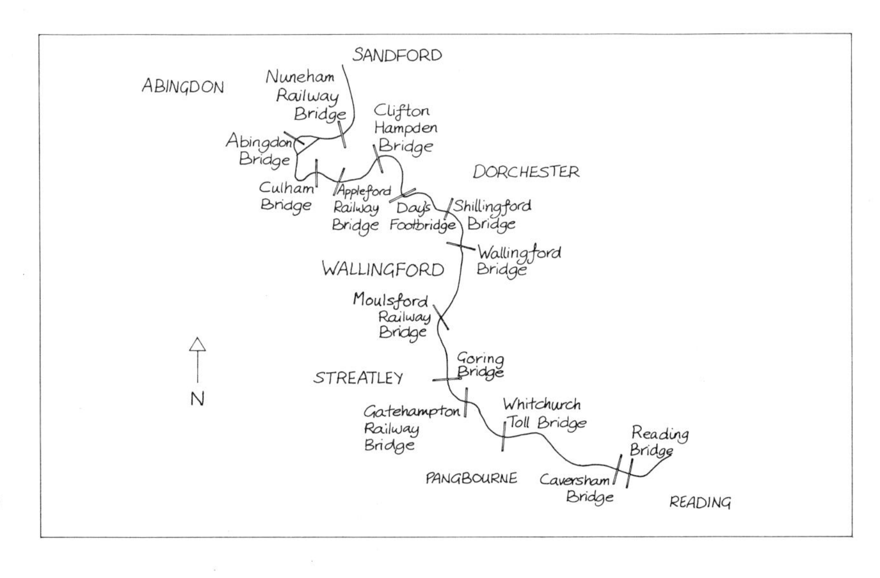

The Thames from Abingdon to Reading. (34 miles).

Although the Thames has flowed past the town of Abingdon for about 1,000 years, this was not the original course between Nuneham and Culham. To find the old channel, one has to look a mile below Nuneham where a small stream separates from the main river and winds its weed-choked way slowly through the countryside until it rejoins the main river near Culham. This stream has been known as 'Swift Ditch' for centuries and must at some time have lived up to its name.

For much of its early history there was no way of crossing the river at Abingdon. Sometimes the river could have been forded upstream at Sandford – a most dangerous crossing because of the currents – or downstream at Culham.

As a result of this isolated position, Abingdon did not develop socially or industrially like the town of Oxford. No doubt it was this isolation which attracted the monks to build the first Abbey here in AD 675. The present course of the river which takes one through Abingdon is claimed to be a 'cut' to the Abbey, constructed by the monks in the tenth century.

The bridge over the Swift Ditch is now closed to traffic both over and under it. The beautiful grey stone and graceful arches are blurred by the spreading weeds of this forgotten backwater.

The river bends sharply and flows on towards the unspoilt pools of Sutton Courtney. The navigation chanel turns into Culham lock cut and avoids these pools. As one negotiates the awkward angle of Culham Bridge, the original bridge comes into view. It is no longer over the main navigational channel, for the water flowing under it comes from the lower end of the pools at Sutton Courtney.

The river passes through flat country, relieved only by the spire of Appleford's church. From Clifton Hampden the river travels in a wide loop. The most distinctive landmarks always in view along this, the Burcot Reach, are the cooling towers of Didcot Power Station, their grey clouds being gradually dispersed by the wind.

As one approaches the ancient town of Dorchester, which lies between the Thame and the Thames, the ugly modern landmark of

Abingdon Church seen through the towpath arch of Abingdon Bridge.

Abingdon Backwater. Upper Reaches Hotel as seen from Abingdon Bridge.

Didcot recedes and the contrasting ancient hills of the Wittenham Clumps or Sinodun appear. The commanding height of these hills over such a wide area made them an important strategic point for the Romans. There is evidence that the earthworks here date from a much earlier time, possibly late Neolithic or Bronze Age. The most important Saxon settlement of the Upper Thames area was Wallingford which is the next major town on the river journey towards London.

Wallingford was a key place in Alfred the Great's system of river forts, and from the times of the Saxon kings until 1270 coins were minted there. Like Oxford, Wallingford was designed and built by the Saxons. It was neatly laid out in a rectangular pattern with its main road running parallel to the river. The pre-historic trackway led from the river crossing to the Downs and the Ridgeway.

Due to its strategic position Wallingford was the first place in the Upper Thames to have a proper bridge – not just a ford or ferry crossing. Well before 1066 there was a wooden Saxon bridge and certainly a more solid bridge in the time of Stephen.

From Wallingford to Cleeve is the longest stretch of river between locks. As one approaches Goring and Streatley, the softly

Culham Bridge – the Lock Cut.

curved and wooded hills carrying the Ridgeway and Icknield Way paths rise above the river. With the Berkshire Downs dipping down to meet the Chilterns at the waterside, this is one the most beautiful parts of the entire river. The London to Bristol railway, engineered by Brunel, follows the valley through the Goring Gap, crossing the river on either side of the Gap itself.

The twin villages of Goring and Streatley are linked by a bridge which replaces a much earlier ford. Travelling downstream the river hugs the hills of the Oxfordshire bank past Pangbourne, famed for its Naval College. The village of Mapledurham, three miles further downstream, nestling around its manor house, appears quite cut off from civilisation. This isolation does not last for long because, after turning a corner, the river flows alongside the railway line once more on its approaches to Reading, the county town of Berkshire.

Early settlers were not attracted by the Thames but by the River Kennet, which flows into the Thames a mile below Caversham lock. Near to this confluence the Kennet broke into several streams that could initally be forded. In Saxon times two major routes (one between London and Bristol, the other between

Oxford and Southampton) are thought to have crossed at Reading, leading to the promotion of the settlement as a trading centre. Yet, at this time, Reading was less important than Wallingford, the latter being a fortified borough permanently garrisoned for the defence of Wessex against the Danes.

The earliest reference to Reading was in 871 in the Anglo-Saxon Chronicle 'In this year the army came into Wessex to Reading'. The Danes entered Reading again in 1006 when they burned the town, together with Wallingford and Cholsey.

Reading Abbey, founded by Henry I in 1121, put the town permanently on the map. The Abbey, however, only lasted until 1539 when it was swept away by Henry VIII. During the Middle Ages the town prospered as a manufacturing centre for wool and leather goods.

With the coming of the railway in the nineteenth century, the area sustained its greatest period of growth – even though barge traffic on both the Kennet and the Thames suffered at the hands of the railways. In this century Reading was one of the first towns to benefit from office and factory relocation from Central London. Today it remains a thriving and bustling, if somewhat drab regional centre.

NUNEHAM RAILWAY BRIDGE

The London to Oxford railway crosses the Thames on both sides of Abingdon. The northern span was first proposed as a timber structure – perhaps to copy one of the many timber designs of Brunel's railway in Cornwall. In the end, though, as an iron bridge Nuneham conforms with the other Thames crossings of the Oxford extension.

The railway was engineered from the south to rise through a short cutting before bursting out onto the two-arch bridge. The line descends to ground level on an embankment meeting the Abingdon branch, trailing in from the left. This branch was originally proposed in 1837, along with the main Didcot-Oxford line, but it was only opened in 1856 after much local opposition.

Nuneham Bridge takes its name from Nuneham House, visible from the railway or river upstream on the right-hand bank.

Nuneham Railway Bridge.

ABINGDON BRIDGE

Crossing the Thames near Abingdon was a perilous undertaking all through the Middle Ages. A ferry took people and animals from the Oxfordshire side of the river to the Abbey, the town and back again. Due to the low-lying fields which were criss-crossed by many small streams, this area was often flooded. In 1316 the abbot and some of his monks were drowned. As a result the monks decided to build a causeway. This is still used as a raised pavement on the approaches to the bridge.

In 1416 bridges were built at both Abingdon and Culham. John Leland (*c.*1530) wrote of the labourer's pay, 'every man had a penny a day which was the best wages, and an extraordinary price in those times'.

The Abbey did not contribute any money to the construction of either bridge. Stone for both bridges was the gift of Sir Peter Bessils from his quarry at Sandford. The labour was paid for by Geoffrey Barbour, a Bristol merchant with Abingdon connections. Money for its upkeep was donated by John Machyns and William and Maud Hales. The bridge, being in three sections, was given three names – Abingdon at the town end, Burford in the middle, and Hales beyond (recalling the Hales' generosity).

The 'Guild of the Brotherhood of Christ', incorporated in 1553,

Abingdon Bridge.

in the reign of Edward VI, undertook the maintenance of the bridge. This was in addition to other buildings like the Almshouses and was for the benefit of the people of Abingdon.

In 1926-27 the bridge was largely rebuilt and the central navigation arch was widened in order to withstand the vast amount of traffic which passes over it on the road to Henley. The modern road authorities (then the county councils of Berkshire and Oxfordshire) claimed that Christ's Hospital should have paid for making the road fit for motor vehicles. They said that as they had looked after the road for centuries they had acquired a legal obligation to do so! After a long dispute the hospital contributed one-tenth of the cost of rebuilding the bridge.

CULHAM BRIDGE

Like Abingdon bridge, Culham bridge was built by the Guild of the Holy Cross in 1416. This bridge crossed the original

Old navigation channel at Culham.

navigation channel of the Thames, which flowed through Sutton Courtney. It was made of stone and comprised five arches. Unfortunately, for river traffic, it was built at a significant angle to the river, which made passage quite difficult.

In 1807, though, the new Culham Lock Cut was opened, reducing the distance from Abingdon. The lock cut rejoined the original course just below the road which crossed the existing bridge. A new single-span stone bridge was thus added to the original one. Although it is a very narrow roadway it carries a considerable volume of traffic from Abingdon to Didcot.

APPLEFORD RAILWAY BRIDGE

When this bridge was being built, in late 1843, it was the centre of an argument between the railway authorities and the Thames Commissioners. The GWR wanted as low a span as possible, for the railway was to approach the river from flat land on both sides. The Commissioners obviously wanted adequate clearance for river traffic and argued that the structure's low height and construction were likely to be most 'injurious to the navigation'.

Looking at the completed structure today it would seem as if the GWR won the day. The single-span girder bridge clears only 13ft above water, making it the third lowest crossing downstream of Oxford.

Appleford Railway Bridge.

CLIFTON HAMPDEN BRIDGE

When Henry Hucks Gibbs became Lord of the Manor of Clifton Hampden in 1842 he decided to improve the estates that he had inherited. Not only did he repair his tenants' cottages, restore the church of St Michaels and All Angels, but he also ordered the construction of a bridge.

Sir George Gilbert Scott, who designed the Albert Memorial

Clifton Hampden Bridge looking towards St Michael's Church.

and St Pancras Station, was chosen to design the bridge. It was constructed of red brick which was made locally in the kilns belonging to Richard Casey. The six ribbed arches are separated by triangular cutwaters, the latter being designed for the safety of pedestrians crossing alongside the traffic. The Gothic style of this bridge typified much of Scott's work in this period.

The bridge was opened in 1867. The toll collector's cottage still stands although the bridge was freed from tolls in October 1946. Nearby stands the famous thatched inn 'The Barley Mow' mentioned in Jerome K. Jerome's book *Three Men in a Boat*.

The bridge required repairs in 1887, but recently suffered more damage. During the long, hot summer of 1983 children were in the habit of jumping off the parapets into the cool waters below. One evening a lorry driver advancing onto the bridge saw some children and thought they were going to jump back into the path of his vehicle. He swerved to avoid them and a tyre burst as he hit the curb, and subsequently the wall of the bridge. Fortunately, he

Temporary fencing on the parapet of Clifton Hampden Bridge in 1983.

managed to brake, otherwise both he and his lorry would have dived into the Thames. The parapet of the bridge had a gap of about 12 ft which was loosely boarded up until the insurance companies agreed on who was to pay for the repair. The damage was eventually put right in April 1984.

FOOTBRIDGE AT DAY'S LOCK

Next to the lock keeper's cottage, just below the Wittenham Clumps, lies a single-span iron footbridge. It was thrown across the Thames in 1870 at a cost of only £50 – probably making it the

Viewed from Day's Lock, the footbridge is dwarfed by the Wittenham Clumps.

cheapest bridge over the river! The original crossing here was a timber swing-bridge with a navigation opening of about 20 ft in width.

SHILLINGFORD BRIDGE

In *The Thames Highway* the author, Fred Thacker, tells the reader that 'the Patent Rolls of 1301' referred to a bridge at this point. It was mentioned in the lease of a fishery extending downwards ' a ponte de Shellingford'. This is the only known historical document which mentions the bridge, so it is by no means conclusive evidence.

There may have been a timber bridge before the present stone crossing, but even this is not very certain. However, there are several historical references to a barge which acted as a ferry. In 1692 Baskerville noted that 'at Shillingford a great barge exists to waft over carts, coaches, horses and men'. The small house which belonged to the ferryman later became the Swan Inn. Today it is the Shillingford Bridge Hotel, a favourite weekend haunt for many visitors, whether staying at the hotel or mooring their boats at the edges of the lawns. On the wall of the main bar there are four old prints of the bridge. The oldest shows the bridge as it was

Shillingford Bridge viewed from the swimming pool of the Shillingford Bridge Hotel.

supposed to look in the eighteenth century. A horse and cart are passing over the bridge set on rickety-looking wooden piers.

The predecessor of the present bridge was built of timber resting on stone piers in 1764, at the instigation of Sir William Blackstone. The Act authorising the building of the bridge was passed in 1763. It was not a very safe structure and was constantly in need of repair. The present stone bridge, consisting of three arches over the river topped by a stone balustrade, was built between 1826-27.

WALLINGFORD BRIDGE

When Fred Thacker was gathering information for his books on the Thames the Rev J.E. Field, vicar of Bensington, wrote to him – 'There is a tradition of a bridge having been erected here soon after the Conquest of the district by the Saxons, but there is little doubt that the earliest stone bridge was that of which a large portion still remains and that anything previous was of timber. Extensive repairs were made in 1530 when part of the Priory Church was bought by the bridgemen and its stones were used for this purpose'.

The Gothic and rounded arch under Wallingford Bridge.

The 'open' season at Wallingford Bridge.

When King Stephen laid siege to Wallingford Castle he built a tower on the Oxfordshire bank at one end of the bridge. Richard, Earl of Cornwall (brother to Henry III), is said to have built the first stone bridge, and the earliest arches of the present bridge date from this time (1231-71). Three of these arches are of thirteenth century stonework but the rest of the bridge has been greatly rebuilt.

During the Civil War Wallingford Castle was held for Charles I. Four arches were removed from the bridge and replaced by wooden drawbridges. These remained in place until 1751 when the arches were reconstructed.

In 1809 the bridge was badly damaged by a flood. Because of the increasing traffic it was decided to widen the road by about four feet and repair some of the damaged arches. The third, fourth and fifth arches were replaced – the fourth arch being known as the Queen's arch. The architects did a very strange thing, though, in designing the widening project. They did not match the new upstream arches with the original arches. The result is that on one side there are some rounded arches with keystones, while the other side still has the original Gothic pointed arches. Even after this widening, today the bridge is restricted to alternate flows of road traffic, controlled by traffic lights.

MOULSFORD RAILWAY BRIDGE

With the opening of the main line west from Reading to Steventon (just beyond Didcot) in June 1840, two examples of Brunel's brick bridgework were constructed, here at Moulsford and two miles downstream beyond Goring.

Moulsford Bridge spans the river over three arches at a skew angle of forty-five degrees. The arches were built in red brick with rusticated stone voussoirs and a stone coping on the parapet.

Unlike the other railway bridges over the Thames that were widened to take four tracks, the two extra lines were accommodated on an almost separate structure upstream. This 'second' bridge, which was opened in 1892, is similar in profile but was constructed entirely of brick without any embellishments. The 'gap' that remains can only be appreciated when travelling under the arch on the river.

Moulsford Railway Bridge. The container train crossing the Thames is on its way from Brentford to Appleford carrying GLC refuse.

GORING BRIDGE

As the Icknield Way and the Ridgeway both drop down to the Thames at Goring it would have seemed a popular choice for a bridge in very early times. However, no bridge was built here before 1837 although the Romans built a raised causeway. For centuries a ferry plied its trade to and fro over the river.

When Goring lock was built the water level rose and both the causeway and the ferry crossing became dangerous. In 1810 the ferry boat overturned and several people drowned. Following this disaster the Thames Commissioners decided that a bridge was required and asked for donations towards the cost. They could not raise enough money initially and it was not until 1837 that Parliament authorised a bridge. It was built of wood with a small toll cottage on the central island.

The timber, particularly the piles and supports, needed constant shoring up and renewal. It was therefore decided to apply to Parliament for authorisation to build a new bridge. This was granted by an Act in 1915. Postponed due to World War 1, the bridge was not actually rebuilt until 1923. The roadway and decking were made of concrete, but the supports and balustrades still remain of oak, as in the nineteenth century.

The bridge between Goring and Streatley showing the wooden supports.

GATEHAMPTON RAILWAY BRIDGE

This bridge is similar in design to the one at Moulsford, although the angle of skew is much less severe. The gaps in the brickwork above track level are there by design. On all bridges and in all tunnels where there is limited clearance between the track and the edge it is mandatory to provide recesses for staff working on the line to keep clear of trains. In this case there are such recesses at regulation intervals, although the men are protected from falling into the river with iron railings.

The bridge is often referred to as Basildon Bridge. This name is taken from the small hamlet of Lower Basildon which is visible from both the river and the railway below the bridge on the Berkshire bank. 'Gatehampton' comes from the two farms on the opposite bank named Lower and Upper Gatehampton farms.

WHITCHURCH BRIDGE

This is the second of the two remaining toll bridges over the river.

Since time immemorial the river could be forded between Whitchurch and Pangbourne at most times of the year. In addition, for centuries there was a ferry which belonged to the Manor of Pangbourne.

In the eighteenth century the shallow water at Whitchurch caused great problems for the increasing barge traffic using the river as its highway. Therefore, in 1787, a pound lock was constructed. This deepened the river so that the ford disappeared. In addition when the river was high and running swiftly in the spring, even the ferry journey between the two villages became dangerous. Early in 1792 the villagers of Pangbourne and Whitchurch petitioned Parliament for a bridge. The bill authorising its building was passed in June 1792.

The bridge was opened to the public in 1793. The toll rate was a halfpenny for a pedestrian, 2d. per wheel and 2d. per horse for each carriage. Barge horses, though, where to pass over the bridge toll-free. The bridge itself was described as a 'neat and light bridge of oak timber with a balustrade on each side'.

By 1850 the foundations of the bridge needed considerable

The painted toll bridge at Whitchurch.

The roadway and toll house over Whitchurch Bridge.

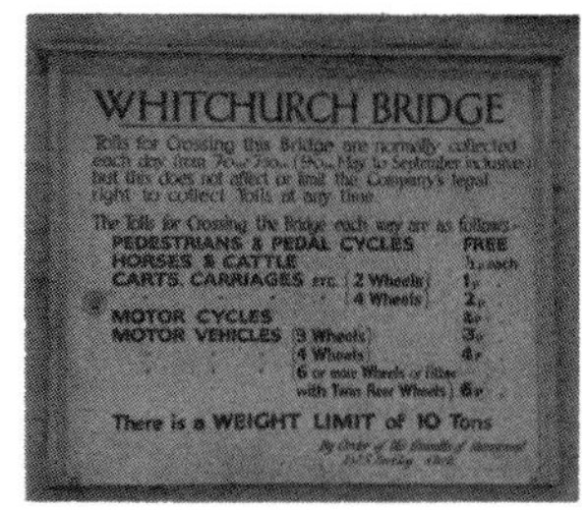

The scale of charges for crossing Whitchurch Bridge.

repair. They had weakened due to erosion caused by the constant flow of water. It was decided that it would be better to renew the bridge entirely and widen the roadway spanning the river. This was completed in 1852.

This timber bridge lasted for about fifty years until the present one was built, in its predecessor's style, but of more modern and longer lasting materials. Today a car passing over it must pay a 3p. toll, but pedestrians and cyclists have free passage.

A Bristol to London train passes through the Goring Gap and over the Thames at Gatehampton.

CAVERSHAM BRIDGE

The earliest reference to this bridge is to be found in the 'close rolls' of 1231. Mention is made of the chapel, dedicated to St Anne, for the use of wayfarers. It refers to the chapel 'on the bridge of Reading ultra Tamisiam, founded partly by the Abbot of Reading and partly by William, Earl Marshall'. At this time the bridge was made of timber and was very narrow.

In the reign of Edward I the first Assizes were held on the bridge. In 1366 the bridge chapel was valued at one hundred shillings and was 'in the gift of Walter, Earl Marshall'.

At about 1480 the people of Reading complained to Edward VI about the abbot's negligence in repairing 'a parte of Causham Bridge with a chapel thereupon to the holy gost'. The chapel was on the right-hand side towards the north of the bridge 'piled in the Foundation for the Rage of the Streame of the Tamise'. (Leland)

During the Civil War an action was fought by the bridge between the Parliamentarians, who were then besieging Reading, and a Royalist force from Oxford under Charles I and Prince Rupert. The bridge was partly dismantled at a cost of £1.9s.7d. When King Charles heard of this he sent an order to the Mayor of Reading that the bridge at Caversham 'be made safe for the army to pass at 8 o'clock next morning'. This was on 1 November 1642. There must have been some delay, however, because there was a written agreement (dated 20 May 1644) with the carpenters for its

Caversham Bridge.

Detail of arch of Caversham Bridge.

'making up' and on 16 July it was 'fitt for any carriages to pass over'.

In 1830 the bridge was renewed, the Berkshire half of wood and iron, the Oxfordshire half of brick and stone. Arthur Humphries said of the bridge – 'The memorable bridge of seven centuries consisted of a brick arch for the tow path, a timber portion under which the river passed and five arches in midstream'.

Although it had been repaired in 1830, by the 1860s it was so dilapidated that the bridge had to be replaced by an entirely new one. It was constructed of iron and opened for traffic on 24 July 1869. Although sturdier than the previous bridge, everyone in Reading complained about its ugly appearance.

As the flow of modern traffic increased in the early part of this century so the bridge proved more and more inadequate. The Reading Corporation Act of 1911 authorised the building of a new bridge. World War 1 delayed its construction and it was not until 1924 that work on the bridge began. It was made of ferro-concrete and spanned the river with two long arches. It was completed in May 1926 but celebrations for the official opening had to be postponed on account of the General Strike. Edward, then Prince of Wales, came to Reading on 25 June 1926 and unveiled the plaque commemorating the opening of the new bridge.

READING BRIDGE

When the Borough Extension Order of 1911 authorised the reconstruction of Caversham Bridge, it recommended that another bridge should be built nearby as a footbridge. When the countil considered the traffic congestion and the fast developing industries in and around Reading, it decided that a substantial road bridge would be more useful than merely a footbridge.

The building of this bridge was also hampered by World War 1. Eventually, the bridge was started at the beginning of 1923. On completion its strength was tested by a spectacular procession of steam rollers before being fully opened on 23 October 1923.

Reading Town Bridge.

4 SONNING TO WINDSOR THE ROYAL RIVER

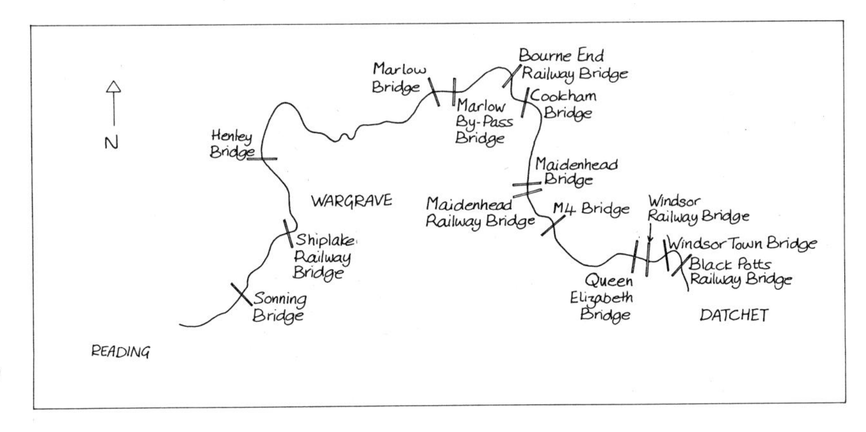

The Thames from Reading to Datchet. (31 miles).

Having left the industrial area of Reading one approaches Sonning Lock and bridge. They form the western boundary of one of the best known and loved stretches of the Thames. From Sonning to Windsor the river winds through beautiful scenery and interesting old riverside towns steeped in the history of England.

According to David Wilson, in his well-documented book 'The Making of the Middle Thames', man has lived in this area since Mesolithic times. Finds of pottery and combs point to the settlements existing in Neolithic times. The Bronze Age left its traces at Bray, as witnessed by plentiful weapons dug up over the years. Burial mounds have been found at Cookham.

With the coming of the Iron Age better and stronger tools were invented. The land could be cultivated and farming developed.

The Romans too, came to the middle Thames. Evidence of their occupation has been found in abundance. Weapons have been

The weight-restricted section of Sonning Bridge.

uncovered which show the likelihood of many clashes between the Romans and the Angle-Saxons.

Windsor Castle owes its origins to William the Conqueror. He built a ring of nine fortresses around London to protect and strengthen his hold on the city. The steep hill at Windsor falling sharply to the river was an ideal place to build a castle. Windsor has seen many monarchs since the eleventh century. Almost every one of them has left his (or her) mark on the castle which has been extended and renovated many times. Today it is much used by the Royal Family at weekends, during Christmas, Easter and Royal Ascot Week. When the Queen is not in residence the public are able to view the state apartments. They hold a tremendous collection of valuables, furniture, paintings and china.

Walking through the cobbled streets of the town, past the seventeenth century Guildhall and the Market Cross down to the river, one passes over the pedestrian bridge into Eton. The famous school was founded in 1440 by Henry VI for seventy boys from poor families. In 1984 there are about 700 pupils, and most of them are from wealthy families. The fees alone are several thousand pounds per year.

Mask of Father Thames on centre arch of Henley Bridge.

When the railways first steamed their way through the countryside, and Brunel built his famous bridge at Maidenhead, the emphasis on river traffic slowly changed. The Thames had always been a working river with barges carrying their goods up and down the length of the river. With the coming of the railways barge traffic declined. A train could accomplish in a few hours a week's journey by barge. The river meanders slowly between locks and the barge hasn't the speed of the train.

As the barge traffic declined so leisure pursuits on the river increased. One event which drew the crowds to the water was the Regatta at Henley. Prince Albert, Queen Victoria's consort, became its patron in 1851. In the forty years before World War I this event was highly popular with the Victorians. Extravagant parties on houseboats were all the rage during Regatta week. Champagne flowed freely. The riverside and all the river craft were festooned with flowers and bunting. Riverside restaurants and ancient inns at Sonning, Henley, Marlow and Maidenhead were filled to capacity.

Today the Royal Regatta still draws its crowds. The extravagance of pre-World War I days has never been matched again. It has often been called 'The Golden Age of the Thames' but it is enjoyed by more people in the 1980s than in the 1890s when only the wealthy could afford to indulge themselves in this way. There are thousands of people who spend their leisure hours by or on the river in an endless variety of pursuits: the towpath walkers; the lock watchers; the fishermen perched like large green mushrooms along the banks – a thousand Toads of Toad Hall under their green umbrellas during the fishing season, sitting patiently, waiting for the bite. There are the river travellers, by hired craft, or their own: sailing boats, steamers, skiffs, narrow boats and motor cruisers, to name but a few. An endless variety of people enjoy the beauty of the river – especially the region from Sonning to Windsor.

This century the riverside towns have prospered and their populations have grown. People have moved farther away from the city centres as transport has developed. One could almost say that the riverside towns have become the suburbs of London. Commuters live in the beautiful countryside of Marlow, Maidenhead and Henley. They build attractive riverside houses and travel each day to work in London or Oxford by road or rail.

SONNING BRIDGE

Although there have been suggestions that there was a bridge here in Saxon times, there appears to be no hard evidence to verify this theory. John Leland (1530) mentions 'Sunning bridge of tymbre' which seems to be the earliest record of the crossing. It was owned by the Ramsbury bishops who were responsible for its repair.

In 1574 Elizabeth I became the owner of the bridge. The bishops exchanged it for other property. Within one year the people of Sonning petitioned the Queen to repair the bridge as its timbers had perished and it was in a dangerous condition. The Queen, though, always cautious with her purse, denied that it was her responsibility. The estimate for timber and labour was a mere £34! After twenty years, when all avenues had been explored and the Queen's Treasurer could find no more excuses the bridge was repaired at a cost of £16. Needless to say, this bargain basement

Sonning Bridge and the White Hart Hotel.

A busy summer's day at Shiplake.

price meant that the job was not well done. Within 10 years the bridge needed repairing again.

In 1719 timber was again used to make the bridge safe. Finally, in the 1770s the part of the bridge that crossed from the White Hart Inn to the centre island was replaced by a brick bridge. This most attractive bridge is best viewed framed by the many willows that surround the crossing.

The two small bridges from the island to the Oxfordshire bank were replaced in 1905 by iron girder bridges which remain to the present day. In 1983 a temporary three-ton weight limit was imposed on this section of the crossing. After a further survey of the decking this limit was raised to five tons in August 1984 – awaiting some unspecified repairs.

SHIPLAKE RAILWAY BRIDGE

A four-and-a-half-mile branch from the GWR main line at Twyford to Henley-on-Thames was authorised in 1847. The line was opened ten years later on 1 June 1857. The original bridge was of timber construction, but when the track was doubled in 1897, it was replaced by the present day iron bridge. That is not strictly accurate, for only half of the 1897 structure remains today. When the line was reduced to a single track again in the early 1970s the downstream side of the twin span was removed. Evidence of this span still remains, however, in the brick abutments and central iron pier which were obviously designed for a two-track crossing.

In its heyday the line saw its greatest activity over the four days of Henley Royal Regatta. The branch was used to capacity with thousands of people travelling in First Class Regatta Specials from Paddington. Even today the branch's shuttle service is augmented on Regatta days by a through train from London.

HENLEY BRIDGE

The earliest record of this bridge was to be found in the Patent Rolls of 1232, in which it said that the bridge keeper was to be given building timber free of toll from the Forest of Windsor. Thacker, in *The Thames Highway*, refers to a legend that this bridge was the 'one which, according to Dion Cassius, the Romans crossed in pursuit of the British, who swam across a lower part of the River'. He does not substantiate it, so perhaps the story should be discounted! Thacker goes on to describe buildings on this bridge similar to those on the old London Bridge. 'In 1354 one of the two granaries leased stood upon it.' It must have been quite a

Henley Bridge.

solid construction to support a granary and houses. Leland refers to it as 'all of tymbre, as moste Parte of the Bridgs be ther about. It was of Stone as the Foundation Shewithe at a low Water'.

This ancient bridge was destroyed in the great flood of 1774. The Act authorising the building of another was passed in 1781, and the present five-arched stone structure was built between 1786–89 at a cost of £10,000. William Hayward, the architect who designed the bridge, died before it was completed. It was his dying wish to be buried under the central arch of the bridge. The people of Henley found the idea somewhat gruesome and so he was buried in the churchyard near the bridge.

The keystone masks of the central arch show the goddess Isis facing upstream and Father Thames facing downstream. The river God's beard is decorated by small fish, whilst bulrushes adorn his hair. Both masks were sculpted by the Hon. Mrs Anne Damer, who lived in Park Place, one of the most beautiful riverside houses near Marsh Lock. She was a cousin of Horace Walpole and perhaps his extravagant praise of Henley bridge was coloured by his feelings for his cousin. He described it as the most beautiful in the world next to the Ponte di Trinita (in Florence). In the 'Bridge Minutes' of 6 May 1785 the authorities only acknowledged one mask when they ordered thanks for the 'very elegant head of Father Thames' which she had sculpted.

The bridge and, in fact, the whole reach of water down to Hambleden lock play an important role each year during the Royal Regatta. All river traffic between the lock and the bridge comes to a standstill while the races are in progress. Girls in elegant dresses carrying parasols and young men in striped blazers and straw boaters wander along the banks clutching champagne, just as they did when the Regatta was first established with the Grand Challenge Cup in 1838. From the centre of the bridge the long, straight course can be seen as far as Regatta Island.

MARLOW BRIDGE

Marlow Suspension bridge, which was opened in 1832, is still one of the most graceful bridges in the Middle Thames valley.

It replaced earlier timber bridges. Although no one has been able to establish when the first bridge spanned this crossing, there have been references to a bridge as far back as 1309. The Letters Patent of that year granted Gilbert, Earl of Gloucester, pontage for four years.

Two centuries later in 1530, Leland refers to a bridge 'of Timbre' at Marlow. In 1642, during the Civil War, the bridge was nearly destroyed by Cromwell's men who were led by Major General Brown.

Marlow Bridge – detail.

Due to the dilapidation of that bridge a new timber one was built and paid for by general subscription. It was opened in 1789. As always, when it came to paying for the work and materials, no one wanted to bear the cost. The counties of Berkshire and Buckinghamshire disclaimed responsibility saying that the bridge was owned by the church! By 1828 the 1789 structure was in a dangerous condition. It was believed that it had been partially made of the timbers of the previous one to cut costs. Repairs seemed impractical so William Tierney Clarke (who had recently finished Hammersmith Suspension Bridge) was asked to submit a design. Marlow bridge consists of a single suspended span of 235 ft with a fine stone portal at each end.

The people of Marlow were so pleased with their new bridge that they built a new church beside it. On the other bank between the bridge and the weir is 'The Compleat Angler' – a riverside hotel with gardens sloping down to the Thames – surely one of the most attractive locations in the Thames Valley.

When, in 1927, it was discovered that the roadway had subsided due to the rusting of the suspension chains, a recommendation to build a new bridge of ferro-concrete was hotly disputed by the people of Marlow. Eventually it was agreed to repair the suspension bridge. In 1965 all the ironwork was replaced by steel. The portals were restored and the bridge repainted. It stands as a monument to Tierney Clarke, who not only designed Hammersmith and Marlow, but also the suspension bridge over the Danube which links Buda with Pest. Although that magnificent bridge was destroyed in World War 2, it was later rebuilt to its original design.

There is one rather unsavoury tale to be told of activities under Marlow Bridge. Many years ago, the landlady of a waterside inn found that bargees would steal food from her larder. In order to revenge herself she baked a 'puppy pie'. This was duly stolen.

The A404 by-passes Marlow, crossing the Thames below Winter Hill.

Secretly she watched the bargees as they ate the pie under the suspension bridge and afterwards she told them what they had just eaten!

A404 MARLOW BY-PASS

The A404 is one of several modern link roads which speed traffic between one busy trunk route and another, allowing heavy traffic to avoid attractive riverside towns. In this instance the town is Marlow and the new bridge not only relieves the centre of Marlow but also the suspension bridge which now has a weight limit and restricted width clearances.

As the A404 bridges the Thames, one can see Marlow lock about half-a-mile to the west and the steeple of the church rising above the intervening trees.

Work on the bridge was started in November 1970 and was completed two years later. It was constructed of pre-stressed concrete and steel and is similar to the one built at Windsor in 1966.

BOURNE END RAILWAY BRIDGE

This crossing was conceived by the Wycombe Railway Company which was authorised to build an initial line from Maidenhead to Wycombe in 1846, with eventual extensions to Oxford and Aylesbury. Construction was delayed by financial problems and the line was not opened until 1 August 1852. These financial difficulties were doubtless the reason for building only a timber bridge to cross the Thames at Bourne End.

Thacker regarded the bridge as a 'terror to navigation'. Evidently, in February 1869 there were many complaints about its dangerous state – barges frequently collided with it and they were sued by the railway company for damage. By this time the company had been amalgamated into the GWR and, as with Appleford Bridge, it seems that the railway was a far more powerful force than the Thames Commissioners.

Nevertheless, the timber structure was replaced by the existing iron bridge in 1893. Today, though, the line no longer extends to High Wycombe. Cut short in 1970 at Bourne End, it now forms part of the Maidenhead to Marlow branch.

Bourne End Railway Bridge.

COOKHAM BRIDGE

This structure, even today, is quite narrow and has no pavement or refuges of safety for pedestrians. The road traffic flows at speed, providing a hazardous crossing for the unwary walker.

Until the nineteenth century a ferry served the two small farming communities on either side of the Thames at Cookham. Mention of the ferry was made in the ministers' accounts of the fourteenth century.

At the Court of Fairs in 1470 a complaint was lodged that the Abbot of Cirencester had neglected the ferry. It was in such a dangerous condition 'whereby people cannot cross'.

In the seventeenth century the ferry was leased to Hugh Cotterell, a local fisherman. By 1840, though, the first wooden bridge was built, by Mr Freebody. It was not a very solid structure and, as with all timber bridges, was destined to be short-lived.

Twenty years later designs were invited for a new bridge. It was constructed of two wrought-iron girders resting on cylindrical iron piers, with attractively patterned wrought-ironwork running along the balustrades. The new bridge was opened to the public in July 1867. The tolls remained until 1947 when they were abolished, although the small octagonal toll-house still stands today.

Cookham Bridge.

Bridge over Cookham Lock Cut.

MAIDENHEAD BRIDGE

The growth of Maidenhead as a flourishing town can be traced back directly to the building of its first wooden bridge in about 1280. A number of ancient weapons found by the riverside suggests that a ferry existed before any bridges were built.

Traffic which had previously passed through Burnham and

Maidenhead Bridge.

Cookham travelled over the new bridge. Trade developed with the building of a new wharf by the bridge.

Pontage was granted repeatedly by the King to maintain and repair the timber structure. The earliest reference to be found was in December 1297, when Edward I made 'a grant at the instance of William de Berford in aid of the Bridge of Maydeheth which is almost broken down'.

The bridge also received money collected by its own 'bridge hermit'. In the Middle Ages it was customary for a hermitage to be established by a bridge. A man (sometimes a priest) would be licensed by the bishop to receive money and gifts from passers by. He would keep just enough for his simplest needs. The remainder would be given to help maintain the bridge.

At Maidenhead there was a hermitage chapel adjoining the bridge which, according to the local church register, was rebuilt in 1423. When it was completed, Richard Ludlow was inducted as the resident hermit. A procession of church leaders, town dignitaries and townsfolk came to the bridge to watch the ceremony and hear Ludlow take an oath renouncing all worldly life and vowing to devote himself to prayer and the bridge.

A charter from James I also gave Maidenhead the right to take three oak trees annually from the Royal Manors of Cookham and Bray for bridge repairs. The Calendar of the State Papers, 11 May

1654, reported that 'the Warden and Bridgemasters of Maidenhead demand three trees and four which they said became due during the war, when they durst not repair the bridge which were several times broken down to prevent the enemy passing'.

Maidenhead Corporation papers show that the old bridge was repaired for the last time in 1750 when £764.9s.2d. was spent on it. The corporation held several meetings to discuss the building of a new bridge. Parliament was petitioned. The bridge Act of 1772 authorised construction. On 23 May 1772 Sir Robert Taylor's plans for a new stone bridge were finally accepted and the foundation stone was laid by the mayor on 19 Oct 1772. The bridge, a structure of thirteen arches took five years to complete. On the occasion of its opening a feast was held at a cost of £43.13s.6d. The total cost of the new bridge was about £19,000 and the money was borrowed upon security of a mortgage on the tolls and rents of bridge land (adjoining the bridge).

Somehow the money collected seemed to find its way to other uses. There were subscriptions to the local races, payments of up to £50 for the Mayor's Feast, expenses for 5 November celebrations and extraordinary amounts for beer and wine on royal birthdays, national victories, coronations and even entertaining preachers! In the Maidenhead Corporation accounts there are 'Expenses at the Sun on the King's birthday 4 June 1772 – £2.6s.0d.' From time to time an honest citizen would protest at a council meeting on hearing of this blatant misuse of bridge funds.

In 1901, a Mr Joseph Taylor of Eton complained. The Corporation ignored his protestations. Mr Taylor then wrote a formal complaint to the Charity Commissioners. An enquiry was held. Mr Taylor declared that if money gained by tolls had been used for other purposes then the bridge should be free of toll. Indeed, the enquiry showed that more than enough money had been collected in tolls to pay all the outstanding debts years before. It required an Act of Parliament to abolish the tolls. Mr Taylor's patience was at an end. He gave notice that on Monday, 8 December 1902 he would cross the bridge in his vehicle without paying the toll. A crowd of about 500 people watched the proceedings with delight. The toll was refused. The toll collector took a cushion from the car and sold it at random for three shillings. He sent the owner of the cushion the following telegram and account:

	s.	d.
Legal Toll	0.	8.
Telegram	1.	10
Registered letter	0.	3.
Balance	0.	3.
TOTAL	3s.	0d.

Mr Taylor received his threepence. The bridge was freed of tolls on 31 October 1903. At midnight a crowd gathered by the bridge. They tore the toll gate from its hinges and carried it to the centre of the bridge. It was ceremonially thrown into the river and the clock on top of the toll house was taken and set up in the Public Library!

MAIDENHEAD RAILWAY BRIDGE

Of all the railway bridges that span the Thames, this is undoubtedly the most famous. Designed by Brunel and completed in 1838, the bridge is believed to include the largest and flattest brick arches in the world. The two semi-elliptical main arches each have a span of 128 ft and a rise of only 24 ft 3 in.

Maidenhead Railway Bridge.

A vision of speed crossing the Sounding Arch of Brunel's Bridge at Maidenhead.

These impressive statistics are the result of the constraints of both railway and waterway which were placed on the design: Local authority requirements necessitated that Brunel design a bridge with only one central pier, while the main arches had to extend over the towpaths. The railway required the minimum rise from ground level up to the river crossing so that the locomotives did not lose speed over rising gradients. Finally, only yellow stock brick was available for construction.

During the period of construction, many critics of the design believed that the bridge would fall down. Indeed, early on, the eastern arch showed signs of movement. (John Fowler, who later widened the bridge, said that its outline became 'perfectly Gothic'). The contractor then admitted that he had eased the wooden centering before the mortar had set.

Continued pressure from his critics forced Brunel to leave the centerings in place still supporting the main arches. But in the autumn of 1839 a violent storm blew this supporting structure away – needless to say leaving the bridge standing firm!

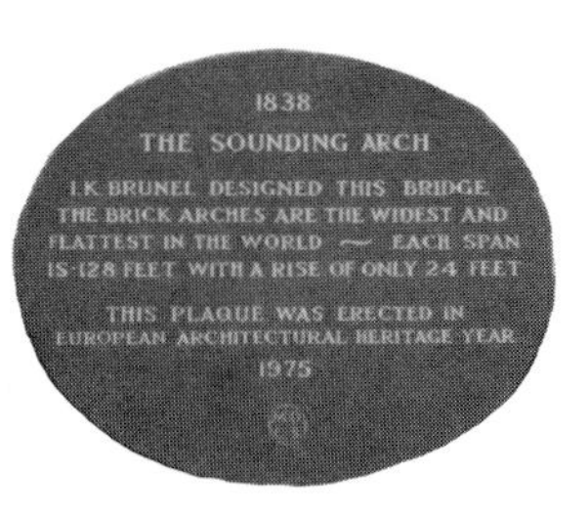

The plaque on Brunel's famous railway bridge.

Between 1890 and 1893 the bridge was widened when the tracks were quadrupled. This work was kept within the limits of the original design, as red brick extensions outside the two 'Sounding Arches'.

In its completed form the bridge has been immortalised in one of J.W.M. Turner's most famous paintings – 'Rain, Steam and Speed'.

M4 MOTORWAY BRIDGE

When a crossing was first proposed here, it was part of a Maidenhead by-pass scheme devised by the Ministry of Transport in 1935. Work began in 1937, but the outbreak of World War 2 stopped construction when only the abutments had been completed.

After the end of the war, road planners had second thoughts on the by-pass. Eventually they suggested that the bridge should form part of the proposed London to South Wales motorway.

A new contract was issued in 1959, with a new single-span design. This consisted of steel plate girders supporting reinforced concrete deck slabs. Its designers were Freeman Fox and Partners, while the famous Horseley Bridge Company were the main

The underside of the M4 bridge showing steel girders which reinforce the concrete deck.

contractors. When it was opened in March 1961 it was named 'The New Thames Bridge, Maidenhead' – as engraved on the commemorative plaque. The name does not appear to have stuck, though – technically the bridge is in Bray, and Maidenhead is some three miles away!

THE QUEEN ELIZABETH BRIDGE (WINDSOR BY-PASS)

In the 1950's both heavy goods vehicles and tourist traffic passing through the narrow streets of Windsor and over the town bridge into Eton increased out of all proportion. The congestion which the traffic caused eventually reached saturation point. The weight and vibration caused by the vehicles undermined the safety of Windsor Town Bridge. The houses along Eton High Street were also suffering structural damage. A new bridge to divert the through traffic became essential. The county bridge engineers, together with Courtney Theobold, designed a new bridge. Work commenced in July 1964 and the bridge was opened in June 1966. It is 82 ft wide and 481 ft long and was named after Queen Elizabeth II.

Windsor By-Pass over the Queen Elizabeth Bridge.

Brunel's Windsor Railway Bridge.

WINDSOR RAILWAY BRIDGE

When Brunel first surveyed the route of the proposed railway from London to Bristol, in 1833, he planned it to pass through the Royal town of Windsor. The Principals of Eton College, however, objected to this move as it would bring the 'corruption' of London within easy reach of the Etonian pupils. So strong was their objection that the GWR was forced to route the line two miles further north through Slough. It is ironic, therefore, that only a few years later Eton College chartered a train to London for its pupils to take part in Queen Victoria's Coronation celebrations.

After ten years of opposition a branch line was built from Slough to Windsor, opening in October 1849. The GWR, like its rivals the London and South Western Railway (L&SWR) had to pay heavily for the privilege of providing a railway. As well as financing the railway itself, the companies had to pay the Crown Commissioners to improve road access to Windsor Castle!

The Thames is crossed on the skew by a wrought iron bowstring girder bridge – a span of 203 ft. The bridge is approached from the Slough side by means of a long brick viaduct, which was originally built of timber. The main structure remains Brunel's oldest surviving wrought-iron bridge.

WINDSOR TOWN BRIDGE

This is a unique bridge across the Thames. There were many bridges which started their lives as mere pedestrian crossings, but this bridge is the only one which after centuries of carrying vehicular traffic has reverted to foot traffic only.

After the opening of the Queen Elizabeth bridge in 1966, all heavy traffic was diverted away from Windsor and Eton town centres to the by-pass. To maintain the town bridge for traffic would have required a large sum of money. So it was decided to close the bridge to vehicles, leaving only foot access between Eton and Windsor. Benches have been placed on the bridge from which tourists can gaze at the Castle, the river traffic and down the main street of Eton.

It is not known how long ago a bridge first existed between Windsor and Eton. There may have been a timber bridge as far back as the time of William the Conqueror, when the castle was built.

The first documentary evidence of a bridge was a reference to tolls collected from traffic passing under the bridge in 1172. In the thirteenth century there was mention of oaks being granted by Henry III in 1236 for repair of the timber. In 1277 pontage was granted to maintain the bridge.

Having destroyed the bridge in 1387, Richard II eventually granted pontage for its repair. In 1649 after the Civil War, a much neglected Windsor Bridge received twenty oaks from Windsor Forest for a substantial repair. This was insufficient because, by 1676, the King was persuaded to grant more timber.

Windsor Town Bridge.

Windsor Castle viewed from the Eton side of Windsor Town Bridge.

The pedestrian way over Windsor Town Bridge.

Even though the bridge was frequently repaired it caused much trouble and Parliament was petitioned for an Act to increase the toll charge, as Windsor Corporation could not afford the upkeep. This was granted in 1736. The extra expenditure did not save it from becoming what Cook described as a tottering rotten old fabric. This was in part due to the lock built at Romney in 1797 which increased the flow of water battering against the bridge foundations.

The new bridge was constructed of three cast-iron arches set into piers of granite. During the time it took to build it, the people of Windsor and Eton had to resort to the ferry which they found very tiresome. They were pleased with their new bridge, but wanted to abolish the tolls. After a long legal wrangle before the Lord Chief Justice and then the House of Lords, the toll gates were finally removed on 1 December 1898.

BLACK POTTS RAILWAY BRIDGE, WINDSOR

On the downstream side of Boveney Lock lies the final bridge of this section of river. Windsor marks a Thames-side boundary in railway terms too, for, having flowed through 'GWR country' up

Black Potts Railway Bridge.

to this point, the L&SWR dominates railway crossings all the way to London.

As was described earlier, it was initially proposed for the GWR to go through Windsor on its way to Bristol, but was held back at Slough to eventually build a branch into Windsor. For the L&SWR, however, the 'Windsor Line' was always viewed as one of a number of suburban branches that radiated from its main artery to Southampton. There was a time, though, when it looked as if the L&SWR would not reach Windsor at all.

When extended from Staines, the line stopped short of Windsor on the Datchet side of the river. Wrangling over the exact location of the terminus delayed opening through to the eventual site at the foot of Castle Hill. This delay also contributed to the late design and construction of the Thames crossing at Black Potts. All this allowed the GWR to claim victory in the race to provide Windsor with railway services – albeit by only two months.

The bridge at Black Potts was designed by the L&SWR engineer, Joseph Locke, and built by Thomas Brassey. It was opened on 1 December 1849, consisting of four 70 ft cast-iron spans, lavishly decorated with Gothic-style ornamental appendages – in tune with the Gothic-style terminus.

The fancy iron work was short-lived, however, for in 1892 the cast-iron arched ribs were found to be fractured. As with its sister bridge down the line at Richmond, the ornate cast-iron girders at Black Potts were replaced by the present-day straight girders of wrought-iron plates.

5 DATCHET TO RICHMOND

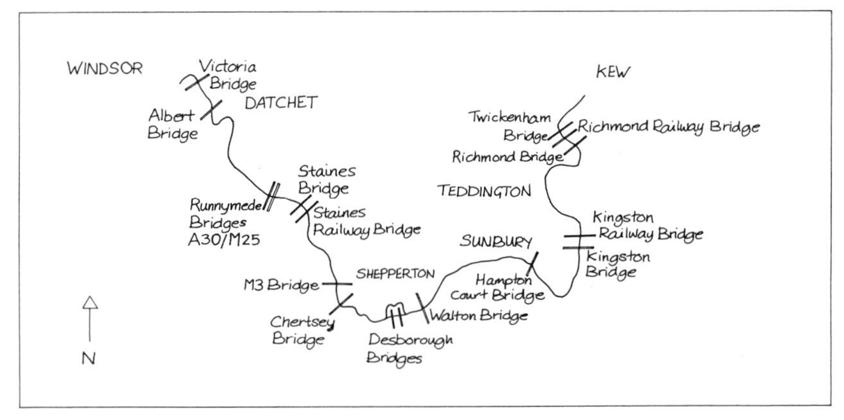

The Thames from Windsor to Kew. (28 miles).

This is the last section of the river before the tideway starts at Teddington. It is not as beautiful as the stretch described in the previous chapter. But it flows past places which are intricately woven into the tapestry of English history.

Having travelled through the royal estates at Windsor and under Black Potts, Victoria and Albert bridges, one passes one of the most famous fields in England – Runnymede. It is a broad expanse of green sloping gently towards the Thames, crowned by Windsor Forest. The oaks from this ancient forest served many a useful task in the repair of old timber bridges spanning the Thames.

Runnymede's greatest claim to fame is, of course, that it is the place where King John signed the Magna Carta in 1215. Whether King John signed and sealed it in the field or on the little Magna

Carta Island is not certain. In more recent times other memorials have been set up at Runnymede. There is the Commonwealth Air Forces memorial for those who died in World War 2, erected in 1953. In 1957 the American Bar Association presented a memorial as a tribute to 'freedom under the Law'.

Lastly, in 1965, Queen Elizabeth II gave the American people one acre of land here on which stands a plain marble slab inscribed very simply 'in memory of John F. Kennedy, born 19 May 1917, President of the United States of America 1961–1963, died by an assassin's hand 1963'.

Leaving the fields and countryside behind one approaches the

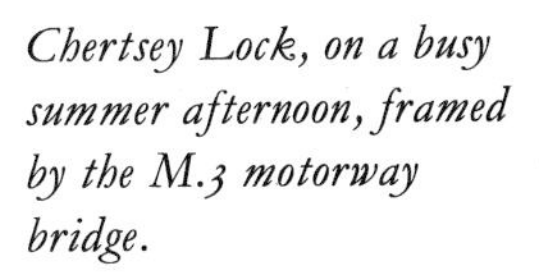

Chertsey Lock, on a busy summer afternoon, framed by the M.3 motorway bridge.

ancient town of Staines – or Stanys as it was called. Today it is a busy industrial and residential town. It was believed to be the site of timber bridges in Roman times, which linked London with the West Country. Staines is mentioned in the Domesday Book in 1086, but there is even earlier documentary evidence in the Patent Rolls of AD 969.

From Staines the river wanders southwards, its path always travelling across the lowest point of the floodplain, past Chertsey. At Weybridge the River Wey flows into the Thames. The natural route of the Thames then describes a loop north to the pretty village of Shepperton, although the Desborough Cut allows traffic to short-cut this winding passage. The ever-widening river then flows towards Sunbury, Hampton and Kingston.

At Hampton the next large slice of English history comes into view. Hampton Court Palace was the riverside home of Cardinal Wolsey, which he started to build in 1514. It was built on a very grandiose scale, with precise formal gardens, as in a French chateau. Cardinal Wolsey, wishing to improve his popularity with Henry VIII, gave the palace to the monarch. After Henry, the palace was not much used by royalty and in 1839 Queen Victoria gave it to the nation. The red brick Tudor building has many interesting features and is a great attraction to visitors, both British and foreign. They come by road and rail, or as Henry VIII himself did, by river. He used an elaborate royal barge, but the tourists come by steamers which run a regular service from Kew and Richmond during the summer months.

Kingston, the next town on the journey into London is indeed an ancient town, being the Coronation site for Saxon Kings. In 1200 it received its first charter from King John and the town's history goes back even further and is related to the strategic importance of the bridge. For centuries Kingston was the first bridge upstream of London Bridge. Control of the bridge was vital to all warring factions, because the bridge opened the road to the west. The Romans, during their occupation, marched their armies over the bridge at Kingston.

A market has been held in the main square since the seventeenth century. If not for the locks built at Teddington, Kingston would have been the beginning of the next chapter. Before the locks controlled the ebb and flow of water from the sea, the tideway extended up river as far as Kingston.

VICTORIA BRIDGE

This crossing was built in 1851. Together with Albert Bridge it serves the function of the old crossing at Datchet, situated between the two present-day bridges. Both are reputed to have been designed by Albert the Prince Consort.

Victoria Bridge on the boundary of the Royal Estate.

ALBERT BRIDGE (DATCHET)

Like Victoria bridge this opened in 1851. One year before, in January 1850, a barge collided with what was called 'the second improvement bridge below Windsor Pound' – evidently referring to the skeleton of Albert bridge. It was described as 'a temporary wooden bridge with scarcely room for barges'

In 1914 there were rumours of the intended rebuilding of this bridge. It remains untouched, though to this day.

RUNNYMEDE BRIDGES (A30 and M25)

There are actually two bridges on this site, both comparatively young. The first – on the upstream side – was opened in

Albert Bridge at Datchet.

The new span at Runnymede carrying the M25.

November 1961. The downstream span is the Thames' newest bridge, having been completed in 1982.

The first bridge was designed as early as 1939 by Sir Edward Lutyens, to form part of the A30 Staines by-pass. As with other bridges designed at this time, construction was delayed by World War 2, and furthermore the A30 was redesigned as a major trunk road rather than simply as a by-pass.

The river is spanned by one arch, 173 ft 6 ins of reinforced concrete and steel. The facing and abutments are of a combination of Portland stone and red brick – as found on Lutyens' earlier work at Hampton Court.

When, in the mid 1970s, the M25 orbital motorway was planned to run parallel with the A30 over the Thames, a second bridge, ten feet downstream, was designed. The two towpath arches and main span of the new structure give a similar profile to the older bridge, although the white cement facing gives it a more modern appearance.

Each bridge carries six lanes of traffic. The original bridge carries the A30 and M25 northbound, while the new span carries southbound traffic.

STAINES BRIDGE

As the Roman road from Verulamium (St Albans) to the encampment at Silchester and the Roman city of Bath leads to the

Staines Bridge.

river at Staines, there must have been a bridge for the armies to cross. Thacker mentions the road 'ad pontes' meaning 'to the bridges'. Therefore, if the two possible Roman bridges are added to the succeeding ones, there have been no less than seven bridges over the Thames at Staines. Presumably the Roman bridges would have been of timber as all the bridges were in those early days. Certainly the third bridge, built here in 1222, was made of wood – and lasted for about 600 years.

From time to time the bridge required repairs. Like so many other Thames bridges it received gifts of oak from the Royal Forest of Windsor to patch up the rotting timbers. Thacker carefully records that 'in 1228 there was a grant of two oaks from Windsor Forest for its repair'. Between 1236–37 further timber was granted on at least four occasions.

Pontage was granted in 1455 to raise sufficient money to pay for the upkeep of the bridge. Tolls were levied for traffic passing both over and under the bridge. These were to be exacted 'for twelve years for the repair of the Great bridge by Stanys and of a causey extending from the bridge to Egham'.

Despite the usual lack of money for the upkeep of the bridge and the various occasions on which the bridge was nearly destroyed by warring factions, it did in fact survive until 1796. In 1791 Parliament authorised the building of a new bridge which was opened five years later. It was built a small distance from the old bridge which was left standing. This was just as well because the new stone bridge had been built on very poor foundations and almost immediately developed cracks. Once again the wooden bridge was repaired and put into use until yet another bridge was built to replace it.

This time it was decided to build an iron bridge. It was opened in 1803, but shortly afterwards it collapsed! The next bridge, also of iron, opened in 1807. In the interim period the old wooden bridge was once again put to use. It was definitely a case of not throwing out the dirty water before one was sure of fresh water, or, in this case, not demolishing the old bridge before the new one was tried and proved!

In any event when this new bridge was opened in 1807, the old wooden structure was finally pulled down. Yet the lifespan of the second iron bridge was also short lived. In 1829 work on yet another bridge was commenced. It was designed by George

Rennie, the son of John Rennie who built London Bridge. It was constructed of white granite and was built on the site of the original bridge – and it still stands today! The opening ceremony was performed by William IV on 23 April 1832, and tolls were finally abolished in 1871.

STAINES RAILWAY BRIDGE

Once the railway to Windsor had been completed railway development south west from London did not stop. Several branches and link lines were built during the 1850s and 1860s. One such route was authorised for the Staines, Wokingham & Woking Junction Railway (soon after absorbed into the L&SWR). The line was proposed to leave the Windsor route at Staines, veering sharply to the left to cross the Thames before continuing on to Ascot and Wokingham.

The bridge itself is formed of wrought-iron plate girders over three 87 ft spans. The two river piers are each formed of three six ft diameter cast-iron cylinders filled with concrete.

Today the two-track bridge looks little different structurally from the day it was opened in 1856. In 1983, however, it did receive a fresh coat of grey-white paint, replacing a somewhat peeling battleship grey.

Staines Railway Bridge.

M3 MOTORWAY BRIDGE

The M3 crosses the Thames in sight of Chertsey lock, the motorway having curved southwards to avoid the village of Laleham and the Queen Mary Reservoir. Although the M3 peters out only a few miles further north, the bridge serves to reduce the congestion which would have occurred if the motorway had ended south of the Thames. The bridge, which was started at the end of 1968, was completed in two-and-a-quarter years. It shows the clean but unremarkable lines one often associates with modern building work. An interesting feature is that the two carriageways not only slope to maintain the general road camber, but are also constructed at different heights.

The bridge is constructed of high-tensile steel. The main span was assembled by three lengths of girders put on barges. These were then towed into position and the sections were then winched up and welded into place.

The lack of history attached to modern motorway bridges and the relative absence of individual 'character' should not cause anyone to ignore their clean, if rather clinical, lines and the benefits which they confer on the travelling public. In this the motive for their construction remains the same as for the railway bridges of the nineteenth century and the pedestrian bridges of earlier times.

CHERTSEY BRIDGE

The first reference to a Thames crossing at Chertsey occurs in 1299. Sibille, the ferrywoman of Chertsey and six men were paid three shillings for taking King Edward, his family and retainers regularly across the river so that they could continue their journey to Kingston. This suggests that there was no bridge during the thirteenth century.

In fact, the first mention of a bridge was by Leland in about 1530, when he wrote that there was 'a goodly bridg of timber newly repaired'. By 1580 the bridge needed substantial repairs again. Everyone agreed that the bridge should be repaired, but no one wanted to pay! Eventually the work was carried out at the least possible cost. The navigation arches were at such an

Painting of Edwardian elegance on gable of house by Chertsey Bridge.

Chertsey Bridge.

awkward angle that river craft found it difficult to shoot the bridge without being dashed by the current against one of the piers.

A new bridge was authorised in 1780. James Paine tendered a design which was accepted. It was to be built of Purbeck Stone laid over piles constructed of fir sunken into iron shoes. Building started in 1780. In 1784 the materials of the old bridge were auctioned. Robert Whitworth, the engineer responsible for demolition of the old bridge, valued them as being worth £240, but they only realised £120. Thacker reported that during that year Whitworth was arrested as a French spy. Whether he was found guilty, thrown into prison or hanged remains a mystery. The bridge, however, was completed in 1785.

Although the bridge had taken five years to build, it had not been well made and the parapets and pedestrian refuges had to be reconstructed. This was not the end of Chertsey's troubles for in 1891 a barge smashed into one of the piers and the centre arch had to be rebuilt.

Since that time the bridge has survived without any major repairs, but it does have a three ton weight limit for vehicular traffic.

THE TWO BRIDGES ACROSS DESBOROUGH CHANNEL

Lord Desborough was Chairman of the Thames Conservancy from 1904–37. During his student days at Balliol College, Oxford, he twice rowed in the Boat Race. It is perhaps from such early beginnings that his strong association with the Thames grew.

Under his Chairmanship the Thames above Oxford was completely modernised and many other improvements were made along other reaches of the river. When a new navigation channel was cut at Weybridge in 1930 it was called the Desborough Cut. The channel is spanned by two single-arched bridges made of stone, with wooden beams forming the balustrades.

A panel on one of the bridges was unveiled by Lord Desborough on 10 July 1935. The work had commenced in 1930 and had been carried out at the joint expense of HM Government, the County Councils of Middlesex and Surrey and the Thames Conservancy itself.

The roadways are quite narrow, with pavements on either side. In 1984 the bridges were restricted to one-way traffic for urgent repairs.

Desborough Channel: the western bridge undergoing repairs.

WALTON BRIDGE

Until the eighteenth century Walton was a small village somewhat isolated from any important routes. Therefore a bridge was not of prime importance to the few people living nearby and a ferry was sufficient for their needs.

During the early part of the eighteenth century more people moved to the area and built fashionable houses. Amongst these wealthier people was a Mr Samuel Dicker, who had returned to England after living on his sugar plantation in Jamaica. He found the lack of a bridge most inconvenient as his business affairs required him to travel to London frequently. But how could he cross the river to reach the Middlesex Road? Travelling by way of Kingston or Chertsey bridges was very time consuming.

On behalf of himself and the growing population he petitioned Parliament in February 1747. The bill became law in June of that year. One condition was laid down, that Mr Dicker should recompense the owner of the ferry, whose income would be severely cut by the existence of a bridge.

Mr Dicker invited architects and engineers to submit plans for the bridge, but he stipulated that the central span should not be

Walton Bridge.

Walton Bridge. The 'temporary' structure is in the foreground.

less than 100 ft wide. This would permit easy passage by the barge traffic. This was not an easy project to undertake in the eighteenth century. Mr Dicker offered a fee of £5,000 and £500 towards the cost of a second bridge at Hampton Court as an inducement. It was several months before a suitable designer could be found.

In London, Westminster bridge was being constructed at the time and one of the designers connected with this famous bridge presented a daring and unusual design. It was of latticed timber on stone piers. Every piece of wood could be individually replaced without disrupting any other. In principle, this was a brilliant idea, but unfortunately the timber did not weather very well.

By 1780 an Act of Parliament authorised a second bridge. It was built by James Paine who was well known because of the other Thames bridges which he had constructed.

The foundations of this bridge were too shallow and in 1859 part of the bridge collapsed. In 1864 a new iron girder bridge was thrown across the Thames. Tolls were abolished in 1870.

Since World War 2, when the bridge was damaged, it has had weight restrictions. In order to avoid disasters a temporary bridge was placed alongside the old bridge until plans are accepted for a new one. As a result of this makeshift bridge built on metal piers Walton offers two entirely different bridge profiles on the same site. No one knows when both structures will be replaced with one new bridge. Until that time the so-called 'temporary bridge' will retain its air of permanence!

HAMPTON COURT BRIDGE

There appears to be some doubt as to whether there were four or five bridges at Hampton Court. Some of the local archives point to five, but there is only clear documentary evidence for four.

The first one was authorised by an Act on 10 November 1747 to link Hampton Court in the County of Middlesex with East Molesey in Surrey. The bridge was built in 1753 under the direction of James Clarke, who was Lord of the Manor of East Molesey and the owner of the ferry. It was made of wood and had seven arches. According to a pen-and-wash drawing in 1754 it was Chinese in appearance, the central arch being decorated by four small pagoda-like structures resting on triangular buttresses.

Hampton Court Bridge with a view to the Palace under the arch.

Attractive and unusual as it was, it proved to be impractical and very soon required extensive repairs.

The second bridge was opened in 1778 and was also made of wood, with narrow arches, creating untold difficulties for navigation. It fared no better than the first and was replaced in 1865 by an iron structure. This was described in the Victoria County History as consisting of eleven arches standing on piles and surmounted by a low parapet. It was painted by many famous artists – Thomas Rowlandson, Peter de Wint, W.B. Cooke and Thomas Shotters Boys.

The arms and crest of T.N. Allen, the proprietor of the third Hampton Court Bridge.

The *Illustrated London News* of September 1866 gave the following description: 'The bridge at Hampton Court lately completed and opened for traffic, is the property of Mr Thomas Allen, having been erected at his expense. Its architectural style is in harmony with the Tudor portion of the neighbouring palace and the bridge has been ornamented with mouldings similar to those on that part of the palace. Foundations are of cast iron cylinders. There are five spans varying from 66 ft to 76 ft and the roadway is 25 ft wide.'

In 1876, amidst cheering crowds, the bridge was freed of toll, to the accompanying sound of the 3rd Surrey Militia Band playing the national anthem.

At Twickenham Library there are newspaper cuttings collected by a Mr Blewett, a local historian, and one of them is entitled 'Tales of a Grandfather'. It describes the following incident: 'The toll master of Hampton Court Bridge seems to have been a determined sort of chap to judge by the action brought against him by Lord Mansfield in June 1767. A poor broom maker, rather than pay the penny toll, appears to have ridden his horse through the river, which was much lower in those days for the water was not then held up at Teddington in the way that it is today. On reaching the bank the toll master seized the horse and later sold it to recover his penny. The trial was a protracted one, but the jury found for the man with the horse and taught the toll master an elementary lesson in civility.'

The fourth bridge was designed by Sir Edwin Lutyens. It is three times as wide as the previous one and has three broad arches to facilitate navigation. To preserve harmony with Hampton Court Palace it was faced with matching red bricks. It was opened by the Prince of Wales (later Duke of Windsor) in 1933.

KINGSTON BRIDGE

Almost every bridge over the Thames has been the scene of some controversy, but Kingston seems to have been the subject of more than most. When it was built of timber it required constant repairs because of the strong tidal force of the water battering against its struts. It was not until the locks of the nineteenth century were constructed that the Thames became tidal only as far as Teddington.

There were never sufficient funds to finance these repairs and too many people disclaimed responsibility. Control of the bridge was fought over by King and Parliament during the Civil War. It was in fact held by Parliament for most of the time except for two short periods when it was under Royalist control.

After much local argument and court proceedings, legislation authorised the building of a stone bridge in the nineteenth century. The roadway carrying the traffic across was too narrow to

Kingston Bridge.

Kingston Bridge at night.

allow two carts to pass each other and the arches were too narrow to allow easy passage for barges! The townspeople complained of the tolls, and when the bridge was made toll free there were even arguments about the celebrations of this happy event.

While there is written evidence of a bridge built at Kingston in the thirteenth century, it is believed that earlier timber bridges existed. Leland, writing in 1539, refers to the contemporary bridge and to an older structure. Thacker wrote that the earliest reference to a bridge that he could find was in 1318 when it was said to be in a dangerous condition. At that time pontage was granted of 2d. on each vessel passing under the bridge carrying goods for more than 100 shillings. In February 1449 a grant of pontage was made for fifty-one years to 'the good men of Kingston' for the repair of the bridge and the causeways.

The next bridge to be constructed at Kingston was also of timber and a manuscript of 1710 relates that 'The Great Wooden Bridge has twenty interstices, two in the middle wide enough for barges. On a post in the middle of the bridge is this inscription in brass, dated 1567, "Robert Hamon Gentleman Bayliffe of Kingstone heretofore has made this Bryge tollfree for Evermore"'. This proved to be somewhat premature as the bridge did not become toll free until March 1870!

Throughout the seventeenth and eighteenth centuries the

bridge was often in a state of disrepair. As the traffic crossing increased, it was found to be increasingly troublesome, especially as the bridge was so narrow (less than 12 ft in width) and very rickety.

In June 1825 Parliament authorised Kingston Corporation to build a new stone bridge close to the position of the old one. It was to be maintained by tolls exacted from pedestrians and horses drawing carts or carriages, but the responsibility for the bridge was handed over to the counties of Surrey and Middlesex.

The bridge was designed by Edward Lapidge. It was made of Portland Stone and its balustrades were of classical Greek style. It is 382 ft long and was at that time 25 ft wide. Each end of the bridge was guarded by a circular toll house. In July 1828 it was declared open by the Duchess of Clarence (who later became Queen Adelaide).

The townspeople who were obliged to cross the bridge each day to work felt most aggrieved at the toll which they had to pay. Eventually, after forty-two years of protests, Kingston Bridge was freed from tolls in March 1870. The Lord Mayor of London and his Sheriffs were invited to the celebration of this event. The Lord Mayor having accepted the invitation, plans for all manner of pomp and circumstance were made. There were to be bands playing, a great banquet and a firework display (arranged by Mr Brock of the famous firework manufacturers). At the last moment news reached Kingston that the Lord Mayor would not attend. The Mayor of Kingston, an Alderman Gould (who had been instrumental in the fight against the tolls), travelled up to the Mansion House and returned in triumph with the Lord Mayor and his retinue. The celebrations lasted for two days and the tolls were finally abolished.

Even though the bridge had been widened to 25 ft, with the ever-increasing volume of traffic it was still not wide enough. In 1906 trams crossing the bridge added to the problems. It was the first Thames bridge across which the trams were permitted to run. This made the roadway even more dangerous for pedestrians, and when a young cyclist was killed crossing the tramlines, it was agreed to widen the bridge. In 1914, at a cost of £66,000, the bridge was widened to 55 ft.

One other quaint tale about this bridge concerns 'the ducking stool'. The *Evening Post* of 27 April 1745 reported that 'last week a woman that keeps the King's Head Alehouse, Kingston, in Surrey, was ordered by the court to be ducked for scolding and was accordingly placed in the chair and ducked in the River Thames in the presence of two or three thousand people'. It appears that this practice has now fallen into disuse!

KINGSTON RAILWAY BRIDGE

Kingston appeared later on the railway map than a town of this size would have suggested. As with earlier instances, the local hostility to railways led the London to Southampton main line to be constructed some miles south of Kingston. Similarly, the town was framed just a few years later to the north by the Windsor line. Kingston was rail linked, initially from the south, with a terminus just short of the Thames.

An act of 1860 authorised the extension of this line across to Hampton Wick and beyond to Richmond. Although designed by J.E. Errington, he died before construction started. Thus W.R. Galbraith took over the design; Thomas Brassey was the contractor – as was the case for several of the L&SWR bridges. The bridge was opened on 1 July 1863. It consists of five 75 ft cast-iron spans, supported by masonry piers.

Kingston Railway Bridge. The power station on the right used to take deliveries of coal from Thames barges.

RICHMOND BRIDGE

There were several references in the household accounts of Henry VIII and his daughters Queen Mary and Queen Elizabeth I to payments made to the keeper of the ferry at Richmond. The ferry consisted of two boats. One was for passengers. The other, a much larger boat, was capable of ferrying horses, small carts and goods.

Richmond Bridge.

Centre arch of Richmond Bridge.

A winter afternoon at Richmond Bridge.

Carriages, though, were too heavy and had to cross over Kingston Bridge.

In the 1770s the ferry was leased to a Mr Windham. He decided that a toll bridge would be a better business proposition than a ferry, and so he petitioned Parliament to pass a bill authorising the building of a wooden bridge.

Much to his surprise there were strong objections on at least three grounds. The people of Richmond did not like the idea of a wooden bridge. Secondly, they protested about the proposed position. Mr Windham wanted to build it from Ferry Hill which was extremely steep and was thought too dangerous for carriages. Thirdly, they objected to the bridge being privately owned and having to pay tolls to the owner, Mr Windham.

It was suggested that a stone bridge should be built at a more convenient place. The idea of a stone bridge was accepted, but its position was to remain the same because the Crown Commissioners would have had to pay for the demolition of 'The Feathers' and several small houses. As they had already paid Mr Windham £4,000 for the remainder of his lease on the ferry, the Commissioners did not feel inclined to spend more on buying out other property.

Footbridge at Richmond half-lock.

James Paine and Kenton Couse were engaged as the architects, and the bridge was completed in 1777.

In Burlington's British Traveller of 1779 it observed that 'the elegant and useful structure consists of thirteen arches, eight of brick and five central arches of stone. They have stone balustrades. The interior part of the three brick arches on the Surrey side are converted to private uses, one as a storehouse, the second a stable and the third a stone mason's workshop. At the entrance to the bridge on the Surrey side is a pyramidial stone which, on two of its sides, are inscribed with the distance from various places'.

The bridge cost £26,000 and tolls were charged until May 1853. In 1937–39 the bridge was widened on the upstream side, the work being clearly visible on the underside of the arches.

RICHMOND RAILWAY BRIDGE

Richmond was connected to Waterloo when the Richmond Company opened a six-mile line to Clapham Junction on 27 July 1846. Windsor, however, was the ultimate goal for the L&SWR

Richmond Railway Bridge under repair.

and in 1847 an Act of Incorporation empowered the Windsor, Staines and South-Western Railway Company to extend the line from Richmond to Staines, Datchet and eventually Windsor.

The river lies less than a quarter-of-a-mile from Richmond station and a cast-iron beam structure was chosen by engineer Joseph Locke to span the Thames. Completed in 1849, the bridge stood for less than sixty years.

In May 1891 a similar cast-iron railway bridge close to Norwood Junction collapsed, leading to considerable questioning of the reliability of the design. Thus, in 1906, the L&SWR authorised the Horseley Bridge Company, under engineer J.W. Jacomb-Hood, to replace the original bridge with a wrought-iron structure. Using part of the old piers and abutments, the new decking and superstructure was completed in 1908.

In 1984 the superstructure again received attention. The main bridge girders and decking were replaced piecemeal over a three month period. The renewed steel structure then received a well-needed repaint.

TWICKENHAM BRIDGE

Although the building of a bridge at this point was first recommended in 1909, it was not opened until 1933. World War 1 and the subsequent shortage of money and materials delayed its construction. Its concrete arches, bronze lamps and balustrades were designed by Maxwell Ayrton. The opening ceremony was performed by Edward, Prince of Wales (later Duke of Windsor). It runs from Chertsey Road, on the Surrey side, through Old Deer Park to join the road to Richmond. Because the bridge's approach cut through Old Deer Park there were many objections to its construction by local residents. It was locally known for a time as 'The Bridge that Nobody wants'!

Twickenham Bridge with the footbridge of Richmond half-lock and weir in the background.

6 TEDDINGTON TO THE TOWER–THE TIDEWAY

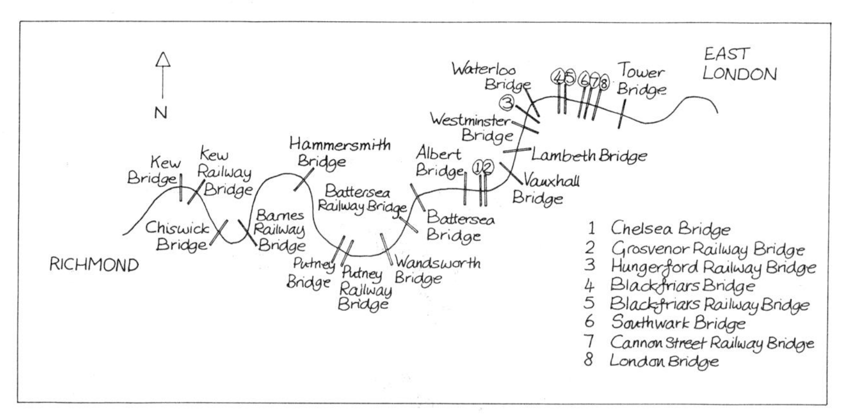

The Thames from Richmond to Tower Bridge. (30 miles).

Looking at the great metropolis of London in 1984 it is difficult to imagine how the city survived with only one bridge over the Thames for 550 years. Peter of Colechurch's bridge, which was completed after thirty-three years of building, had a monopoly over the London river until Westminster Bridge was opened in 1750. A temporary timber structure spanned the river at Putney in 1729. But for over 500 years there was no bridge between London Bridge and Kingston Bridge. In 1984 there are twenty-six bridges over the tidal reach of the Thames. There are road and rail crossings over the water and tunnel crossings beneath the surface.

Why was there only one bridge? The reason was the strong opposition to more bridges from the City Corporation and the ferrymen. Taking the latter first, the explanation is simple. The several thousand ferrymen would lose their livelihoods. For centuries they carried people, carts and animals over the water

Reflections at Kew Bridge.

between the banks. The more bridges built across the river, the fewer ferries would be required.

People would sooner pay a bridge toll than pay a ferryman. Ferries were a tedious and sometimes dangerous way of crossing. The second group of people to object to additional bridges was the City Corporation. Since it was granted its charter by Richard I in 1193, it has always played a dominant role in the growth of London. Strong rivalry developed between the City Fathers, who controlled finance and trade, and Westminster, which was the political centre of England. The City Fathers wanted to contain London within the city boundaries. The corporation felt that the population would be easier to control if it did not spread outwards along the banks. Further bridges would merely encourage people to move away from the central area.

The result was that London became a very crowded city. Everyone wanted to live and work as close as possible to London Bridge. The streets were narrow. The houses were tall and dark. The density of the buildings took away much of the light. Fresh water supplies and sewage disposal posed major problems. Polluted air hung over the city like smoke in a windowless room. The freshest air was on London Bridge, which made its houses much sought after.

Fisherman beneath Kew Railway Bridge.

When a scheme to build a bridge between Lambeth and Westminster was proposed, the City Corporation offered Charles I a huge loan of £100,000 if he would refuse his assent. As he was perpetually short of money, he accepted the offer. The plans for the bridge were shelved. The plague of 1665, followed by the Great Fire of 1666, spurred Londoners on to move further afield. People spread along the northern bank to Westminster, and along the southern bank from Southwark to Lambeth.

For the next ninety years the people of London struggled on with only one bridge. River traffic still formed a major link in communications. Parliament was lobbied continuously to pass an Act authorising another bridge. Finally, assent was given. The bridge was built. In 1750 Westminster, London's second bridge, was opened.

Having overcome the opposition, a spate of bridge building began over London's river. Kew Bridge was opened in 1758; Blackfriars was built in 1769. Battersea's timber bridge followed two years later. The bridge at Richmond was opened in 1777. Then there was a pause whilst Britain was engaged in fighting the French. Vauxhall Bridge was opened in 1816, followed by Waterloo (named in honour of the Duke of Wellington's victory) in 1817. The little used Southwark Bridge was opened in 1819, and 1827 saw Tierney Clarke's new type of bridge at Hammersmith – the first suspension bridge over the Thames.

The next phase of bridge building came in the mid-nineteenth century after London Bridge had been rebuilt by Rennie between 1825–31. Between 1858 and the end of the century five bridges were built on new sites. The first of these was Chelsea in 1858, followed by Lambeth in 1862. Albert and Wandsworth both opened in 1873. The first three bridges were built on the suspension principle, following the success of Hammersmith.

All these new nineteenth century bridges spanned the river upstream of London Bridge. As the Port of London docks developed more people lived and worked on this eastern side of London. It became increasingly apparent that some other river crossing was urgently required. Tunnels were suggested as an

One of five elaborate cast-iron screens that top the piers of Kew Railway Bridge.

alternative to bridges. After many engineering problems, Brunel's first tunnel was opened in 1843. What was needed, however, was another bridge.

The main problem was how to construct a bridge tall enough to allow free passage of tall ships to the Pool of London. Eventually Tower Bridge was designed, built and finally opened in 1886. At first its bascules were lifted frequently. In the later part of the twentieth century the drawbridge is seldom raised, not more than three times a week. On special occcasions it is raised as an accolade: when the Queen returned on Brittania after a succesful foreign tour; or when Sir Francis Chichester sailed his ship Gypsy Moth up to London on his return from his lone journey around the world.

Four London bridges are maintained by the ancient institution of the Bridge House Estates Committee of the City of London Corporation. They are London, Blackfriars, Southwark and Tower. The Bridge House Estates own a large amount of property and funds. It is from these that the committee is able to take responsibility for the costs of all four bridges at no expense to the ratepayer or taxpayer.

Since Tower Bridge was opened in 1886, no new bridge has been built over the Thames in London, although several have been repaired or reconstructed. To ease the evergrowing congestion of the spreading suburbs of London, a new bridge is contemplated further east at Thamesmead. Construction is thought likely to commence in 1985/86.

In the late 1960's and 1970s the Port of London declined. The requirements of an efficient modern port had changed considerably. The method of shipping by container and the increasingly bulkier cargo carriers required deeper water than the Port of London Docks could offer. A modern port with better facilities was needed if London was to compete with its European counterparts in Rotterdam, Antwerp and Hamburg. Tilbury Docks, nearer the estuary, seemed to fit the bill. It was a safe, deep water port, with sufficient land around it for development.

The decline of the London docks left once busy buildings looking like empty shells. The Port of London Authority decided that these disused areas must be redeveloped. No longer will they have the derelict and neglected look of an ancient slum. Imaginative architecture is transforming the docks into some of the most attractive riverside residences and marinas in London. St Katherine's Dock was the first to have such a face lift in the 1970's. Adjacent to Tower Bridge, it now sports a yacht marina, the Tower Hotel, a trading and conference centre and the head office of the Port of London Authority itself.

KEW BRIDGE

There have been three bridges connecting Kew with Brentford. Before the first bridge was built in 1759, a ferry took people and horses across the Thames at this point. From 1659, until the opening of the bridge, the ferry was owned by the Tunstall family.

Middlesex coat of arms on Kew Bridge.

Robert Tunstall petitioned Parliament for a bridge at Kew.

In 1757 Parliament authorised the building of a bridge. The work was carried out by John Barnard who had worked on Westminster Bridge. It was mainly constructed of timber piers to support the side arches, and the central span of 50 ft. Two days before the official opening the Prince (later George III) and the Princess of Wales, passed over the bridge and made a present of £200 to Mr Tunstall and 20 guineas to the workmen. On the first day 3,000 people crossed the bridge, paying the toll of 1d. each.

By 1774 the bridge required extensive repair. The scour of the water had done irreparable damage to the wooden piers. Robert Tunstall obtained authority to replace the timber structure with a new one of stone. This was designed and built of Portland stone by Jame Paine. It was opened in September 1789 by George III followed by a long procession of carriages. In 1819 the Tunstall family sold the bridge to a Mr Robinson. In 1845 he sought leave to build a landing place for steamboat passengers. He stated that 'during the high tides the Steam Boats occasionally land their passengers on the Kew Side, and great rioting and disorder having been occasioned by the Barge Carters drawing their towing lines over the Steam Boats, and I myself have on one occasion seen the Funnel pulled down and great alarm occasioned to the passengers in consequence'.

His petition was allowed in April 1848. But in June he wrote 'owing to the absence of a pier at Kew, 1,200 or 1,400 persons in one day landed at Strand on the Green, and had to walk half-a-mile and across the toll bridge to the Botanical Gardens'.

In 1873 Mr Robinson sold the bridge to the Metropolitan Board of Works, who in turn sold it to the Surrey and Middlesex County Councils. As the traffic between the two counties had increased since the bridge was built, they decided to widen it. Sir John Wolfe Barry, who was asked to make a report on this project, informed them that the bridge should be rebuilt for both safety and economic reasons. The new bridge was opened by Edward VII in May 1903. In the centre of the eastern balustrade a plaque commemorates this event. The bridge was named after the king, but if anyone should ask for directions to the Edward VII Bridge no one would know it by that name. It has always, in the past, been known as Kew Bridge, and surely will continue to be so in the future.

KEW RAILWAY BRIDGE

The opening, in 1869, of this crossing allowed many suburban services to operate from North and West London into Richmond.

Kew Railway Bridge.

Over the years services have been operated by six different railway companies from different parts of the capital. Today London Transport District Line trains and the Broad Street service of London Midland Region operates over the bridge, although the structure itself is still owned and maintained by the Southern Region of BR. From 1986, with the electrification through Stratford to North Woolwich, the BR North London line service will become fully orbital – terminating on the north bank of the Thames some 22 miles down river from Kew.

When the bridge was built, engineer W.R. Galbraith's design came in for a fair amount of criticism. Although a fairly standard iron lattice girder bridge, it was built with some unusual decorative details. The iron piers which support the five spans have three stages – a cylindrical base, a drum with four engaging columns and then, above track level, a tabernacle with an arched roof carried on a pair of columns, framing an elaborate cast-iron screen.

Despite these Gothic capitals, many shared Thacker's view that the bridge 'successfully disfigures the attractive riverside hamlet of Strand on the Green'. For today's environment, though, it can be argued that the crossing is one of the more elegant iron lattice bridges and, in its setting, is certainly more pleasing to the eye than many more modern structures.

CHISWICK BRIDGE

The authorising Act was passed in 1928 for the construction of this bridge. Building Chiswick Bridge was part of a major rebuilding and improvement scheme for the Great Chertsey Road. Twickenham Bridge and Hampton Court Bridge were part of the scheme and were all opened by the Prince of Wales on 3 July 1933.

Chiswick Bridge was designed by Sir Herbert Baker, with Alfred Dryland as the engineer. The bridge is very similar to the crossing at Twickenham. Both were made of ferro-concrete and faced with Portland stone. The bridge is 607 ft long and 70 ft wide between the parapets. Its greatest claim to fame is that it is virtually on the finishing point of the annual Oxford and Cambridge University Boat Race.

Chiswick Bridge.

BARNES RAILWAY BRIDGE

The same Act that authorised the construction of the railway across the river at Richmond to Windsor allowed the Windsor, Staines and South Western to build a $7\frac{1}{4}$-mile line from Barnes to Feltham. For its Thames crossing, Joseph Locke and Thomas Brassey designed and built a three-arch bridge of cast-iron. The line opened on 22 August 1849 and was later to provide a useful by-pass for through passenger and freight traffic avoiding the congested line through Richmond.

Indeed, it was this increase in traffic which prompted the railway to strengthen the bridge in 1894/5. Edward Andrews designed the new brick abutment and piers together with the wrought-iron bowstring girders which make this bridge so distinctive. He also had a footbridge constructed on the downstream side which still remains today.

The bridge is a prominent landmark at the closing stages of the

Barnes Railway Bridge viewed from Barnes Railway Station.

Boat Race. Today its only significance is that the footbridge is closed to pedestrians during the race. But in former years the railway organised special services to stop on the bridge at the time of the race, allowing spectators to view the classic race from a comfortable grandstand.

The Railway Gazette of 29 March 1935 notes that three trains were run that year from Waterloo, leaving the terminus at 1.50 pm, 1.59 pm and 2.06 pm, at a return fare of half-a-guinea 'inclusive of entertainment tax'. The Gazette muses that this considerable premium over regular fares helped to pay for the extra cleaning needed after the privileged passengers had stood on the seats to obtain a clear view!

HAMMERSMITH BRIDGE

The original bridge was authorised by an Act of Parliament in 1824. It was the first suspension bridge to span the Thames and was designed by William Tierney Clarke, an engineer to the West Middlesex Water Company. Following his success at Hammersmith he built another suspension bridge at Marlow and subsequently one in Budapest. Tierney Clarke lived in Ham-

Southern pier of Hammersmith Bridge undergoing repair in 1984.

mersmith and was buried in the local parish church. On his memorial stone there is a representation of Hammersmith Bridge.

His bridge was built of stone on two brick piers, above which there were two towers with arched entrances in Tuscan style. From the towers eight wrought-iron chains held the bridge in position. Building it on the suspension principle provided a wider area under the bridge than the usual arched span. Nevertheless, since the deck was slung low across the water it impeded the passage of navigation except at low tide. Sometimes, steamers with high funnels would have to wait for the water to recede before passing safely under the bridge. To control the traffic flow on the bridge, octagonal toll houses were erected at both ends of the structure.

By 1870 there were fears that the deck was not strong enough to carry the increasing and heavier vehicle traffic. A temporary bridge was built in 1884. Sir Joseph Bazalgette designed a new bridge which still stands today. It was more elaborate than the first one and Sir Nikolaus Pevsner criticised it as a 'Suspension Bridge with atrocious, partly gilt iron pylons crowned by little Frenchy pavillion tops, and with elephantine ornaments at the approaches'. It is, of course, a matter of individual taste!

In more recent times there has been concern over the strength of this particular crossing. In1975–76 the road decking was completely renewed, a long process that caused much disruption to local traffic. Then in late 1983 two of the vertical cables 'failed'. The bridge was closed to all but foot traffic for a number of weeks while engineers investigated the fault. In this time there was much comment in the papers that Hammersmith Bridge was in danger of falling down, and that several other Thames bridges were in a perilous state!

After many tests it was decided to open the bridge to light traffic (and buses). The final test of this involved moving a number of double-deck buses, nose to tail, onto the bridge to check for movement. It turned out to be months, though, until the problem was totally resolved and the scaffolding on the southern pier was removed.

PUTNEY BRIDGE

For centuries there was no bridge over the Thames between Kingston and London Bridge. At various places, including Putney, the north and south banks of the Thames were linked by ferries. As Putney and Fulham's population grew, crossing the river by ferry became more and more tiresome. The people wanted a bridge. Despite much opposition from the ferry owners and the City Corporation, Sir Robert Walpole and his supporters successfully petitioned Parliament for a bridge.

The Act of 1726 permitted the building of a bridge, provided that the owners of the ferry were fully compensated. This in itself

Putney Bridge.

led to an amusing result. Initially several sums of money were paid to certain inhabitants of Putney and Fulham, such as widows of watermen and churchwardens. However, as the manors of both places were jointly owned by the Bishop of London and the Duchess of Marlborough, they were given free passage over the bridge, but the King (and all his other subjects) had to pay the toll!

The bridge had twenty-six arches and the painter J.W.M. Turner thought it so attractive that he made it the subject of one of his famous riverscapes.

As with all timber bridges it required considerable maintenance and, when in 1870 a barge damaged the three central sections, they were replaced by an iron girder of 70 ft. In 1879 the Metropolitan Board of Works bought the bridge and, within a few months, they proposed the building of a new bridge.

It was designed by Sir Joseph Bazalgette, but was very different to his work upriver at Hammersmith. Concrete and granite were used to make a really solid bridge, 700 ft in length and 43 ft wide. It no longer suffers the consequences of barge traffic hitting rickety piers, but in 1984 when the Cambridge University boat collided with a barge just up-stream of the bridge the restricted vision when coming through under the bridge was given as the cause of the accident!

PUTNEY RAILWAY BRIDGE

There has been a railway at the north end of the present bridge since March 1880. The station there incorporated a footway to a new low-water pier where passengers could make connections to pleasure steamers.

There had been Parliamentary powers to develop the line south across the river for some time, but it was by an Act of 25 June 1886 that the L&SWR was able to start building a route connecting Putney Bridge Station with Wimbledon. Work started in March 1887, Lucas & Aird being awarded the contract for construction. William Jacomb, the resident L&SWR engineer, designed the bridge as an eight-span wrought-iron girder structure. Never given a name, it was dubbed 'The Iron Bridge' by the locals.

Train services commenced on 3 June 1889, and a footway on the downstream side was opened the following month. Today, the bridge is part of the District Line service from Wimbledon via Earl's Court to Edgware Road and Upminster.

Putney Railway Bridge.

WANDSWORTH BRIDGE

Due to the proposed construction of the Thames Embankment and the uncertainty of its path, the building of Wandsworth Bridge was postponed several times. The first Act authorising the bridge was passed in June 1864.

Two further acts lapsed before the bridge was finally built in 1873. It was a latticed wrought-iron structure, designed by Julian Tolme, with cross girders supporting the timber roadway. When it was first opened a toll was charged, but the bridge was freed from tolls in 1880.

The London County Council were owners of the bridge by this time. They decided that because of the rusting ironwork, the narrow road deck and the general neglected condition of the bridge, it needed to be replaced. Matters moved slowly and it was not until 1935 that the Council gave the consent for a new bridge. Sir Pierson Frank designed the new structure, consisting of a 200 ft central span, formed of seven high-tensile steel girders, carried on fixed bearings at the piers. It is supported by two 90 ft side spans constructed to a similar principle. The bridge was opened, after some delay, in September 1940.

Wandsworth Bridge.

BATTERSEA RAILWAY BRIDGE

Although the only north-south through-route in London, this bridge is traversed by the sparsest of all British Rail services across the river. The bridge was opened on 2 March 1863 as part of the West London Extension Railway. This railway was authorised four years previously to connect the main lines radiating to the north (out of Paddington and Euston) with the southern lines of Waterloo and Victoria at Clapham Junction.

Co-operation between competing railway companies reached new heights with the construction of this railway! It was jointly owned by the London & North Western and the Great Western (one-third each) and the London & South Western and the London, Brighton & South Coast (one-sixth each). Moreover, the track was laid for both standard guage and the GWR's broad guage.

It has to be said that the potential offered by this much-linked line has never been realised. The multiplicity of connections led to considerable development of freight traffic. Yet passenger services only commenced in 1904, and even then did not attain the proportions that the potential suggested. Through services

Battersea Railway Bridge.

between the north and south ceased in 1939, leaving the peak-hour shuttle between Kensington and Clapham Junction as the sole surviving local service.

The line formed part of the proposed Channel Tunnel route to a revitalised Kensington Olympia – but this, of course, has yet to come to fruition. In 1984 the GLC undertook a feasability study for recommencing local services up to Willesden – but this too has yet to be actioned.

Long distance passenger traffic, however, has returned to this Thames crossing. In 1979 BR instituted two daily return Manchester-Brighton Inter-City services – to serve the growing Gatwick Airport. From Birmingham these trains follow the Western Region main line down the Thames Valley to Acton and then diverge onto the West London line across to Battersea. In 1984 the service was increased to three daily services. When this five-arch structure is reached, passengers from north of Oxford will have crossed the Thames no less than seven times. This particular crossing has the distinction of being the Thames' slowest railway crossing – all trains are restricted to 15 mph so there is plenty of time to view this fairly industrial reach of the river!

BATTERSEA BRIDGE

The first bridge at Battersea was built of timber because Earl Spencer, who owned the ferry, could not raise sufficient money to span the river with stone. The Act authorising the building of the bridge was passed in 1766. It was built by John Phillips and Henry Holland.

Although it was the subject of a romantic painting by James Whistler, entitled 'Nocturne', it was a most unpopular bridge. It had nineteen narrow spans and barges and other river traffic often collided with the piers and many people were drowned.

The piers and the wooden railings along the roadway required frequent repair until it looked like a piece of much darned material with little of the original fabric left!

When the Metropolitan Board of Works bought the bridge they inspected it and found it so dilapidated that it urgently required

Houseboats by Battersea Bridge.

replacement. Sir Joseph Bazalgette designed the new five-span bridge. The work was carried out by John Mowlem & Company. They fixed a temporary crossing until the new approach roads and bridge were built. The opening ceremony was performed by the Earl of Rosebery on 21 May 1890.

ALBERT BRIDGE

Although the Act of Parliament authorised the Albert Bridge Company to construct this bridge in 1864, it was not completed and opened to traffic until September 1873. The delay was caused partly by the objections from other toll-bridge owners and partly because of indecision affecting the building of the Embankment along this reach of the river.

The rigid suspension bridge was designed by Rowland Ordish in what Philip Howard describes as 'fussy Victorian grand manner, with intricate cats-cradles of cable and curious little pagodas that crown the supports'.

The Albert Bridge Company was also made responsible for the repair and maintenance of Battersea Bridge. Having spent all the

The northern pier of the Albert Bridge.

available capital on building Albert Bridge they could not afford to maintain either one on the money raised by toll charges. In 1878 the bridges were bought by the Metropolitan Board of Works, and freed from tolls in 1879.

In 1884 Sir Joseph Bazalgette (who designed the London Embankment as well as some of the bridges) inspected Albert Bridge and found metal girders and cables rusting and corroded. The bridge was strengthened and modernised, and a five ton weight limit imposed on vehicles. For the next sixty years the bridge was well maintained.

Albert Bridge notice to soldiers.

After World War 2 the London County Council decided, on inspection, to pull the bridge down. There was a tremendous outcry by the Chelsea conservationists, led by Sir John Betjeman, who later became Poet Laureate. They overcame the town planners and the bridge was saved. To strengthen it against the heavy demands of London traffic two concrete piers were placed under the main span to give it extra support, and a new lighter deck was laid. The previous weight limit was further reduced to two tons. On each end of the bridge there is a notice directed towards the inhabitants of Chelsea Barracks, requesting that all marching soldiers should break step when crossing the bridge!

In 1983 the bridge was repainted. The colours are like those on a delicately iced cake; pinks and yellows, pale green and light blue. At night the suspension chains are lit like an intricate necklace traced across the darkness, reflecting onto the water below. To have replaced this elegant Victorian structure with a modern concrete span would indeed have been a loss to the visible history of bridges over the Thames.

CHELSEA BRIDGE

After Westminster Bridge was built in 1750, the population of London rapidly spread both eastwards and westwards along the banks of the Thames. By the middle of the nineteenth century Fulham, on the northern bank, and Battersea, on the south bank, had large and expanding populations. A bridge between the two suburbs became essential.

An Act authorising the building of a bridge was passed in 1846. The engineer, Thomas Page, designed a suspension bridge. Work started in 1851 and the bridge was completed in 1858, when it was opened by the Prince of Wales.

The Act of 1846 permitted tolls to be charged until such time as the building costs had been re-couped. In 1879 the tolls were abolished.

Although the bridge was strengthened with additional chains by Sir Joseph Bazalgette in 1880, only forty years later a new bridge was recommended. In 1935 Sir Pierson Frank, in conjunction with Rendel, Tritton & Palmer started work on a new

Cadogan Pier and Albert Bridge from the Chelsea Embankment.

Night reflections of Albert Bridge with Chelsea Bridge in the background.

A golden galleon decorating lampstand on Chelsea Suspension Bridge.

Early morning at Cadogan Pier looking towards Chelsea Suspension Bridge.

suspension bridge. Stronger foundations, set inside granite, were bored into the riverbed. The new roadway was suspended by using thirty-seven galvanised steel wires.

As a large amount of Douglas Fir from British Columbia was used in laying the roadway, it was decided that Mr Mackenzie King, the Prime Minister of Canada, should open the bridge, which he did in May 1937.

GROSVENOR RAILWAY BRIDGE

This bridge has the distinction of being the first railway bridge across the Thames in London. Opened in June 1860, the London Brighton & South Coast Railway (LB&SCR) was able to extend its operations into the West End at Victoria Station. The engineer, John Fowler, was under instruction from the Thames Conservators to design a bridge where the piers conformed to Chelsea Bridge, some 150 yards upstream.

Being the first into London, the LB&SCR was able to lease out track and station space to several other companies – even allowing the broad-guage GWR into Victoria. The London Chatham & Dover Railway had been granted running powers into Victoria and, when the two companies got their heads together, decided to construct a second bridge in order to accommodate more tracks.

The new 100 ft wide bridge was designed by Sir Charles Fox to

Grosvenor Railway Bridge with Chelsea and Albert Bridges in the background.

match the existing bridge. Opened in 1866, by the turn of the century it also became insufficient to the needs of the railway. So a third bridge was added in 1907 to bring the tracks up to ten.

In the 1960s the complete bridge was replaced piecemeal, to bring the whole structure up to date. Today it is one of the few railway bridges in London that has not had to be rationalised, for Victoria remains one of the busiest terminals in the capital.

VAUXHALL BRIDGE

The Act authorising the building of this bridge was passed in 1809, allowing one more bridge to ease the congestion between north and south of the Thames.

It was the first iron bridge to cross the Thames and was originally named in honour of the Prince Regent. It was 36 ft wide and 809 ft long. Its nine cast-iron spans stood in deeply embedded stone, faced in granite. James Walker, the engineer and designer, was determined that the bridge should be sufficiently strong and solid to withstand the damaging ebb and flow of the tidal Thames.

The bridge was opened to the public in June 1816. From then until 1879 it was a toll bridge, with tolls varying from 1d. for

Vauxhall Bridge with the statue of a muse looking downstream.

Dawn at Vauxhall Bridge.

pedestrians to 1s. 6d. for heavy vehicles. Soon after the Metropolitan Board of Works bought the bridge in 1879 it was made toll-free. The Board inspected their new purchase and found that the piers were in a dangerous condition. Repairing the bridge was too expensive and building a new one was delayed.

In 1895 the new owners, the London County Council petitioned Parliament for permission to construct a new bridge. Authorisation was granted that year. A temporary wooden bridge was thrown across the water and work on the new bridge started in 1904. It was opened in May 1906 by the Prince of Wales.

The Chief Engineer to the London County Council designed the present Vauxhall Bridge. Once again Vauxhall is a unique bridge in that its piers are decorated by larger than life-size statues sculpted by F.W. Pomerey and A. Drury. Looking towards Westminster, the figures represent Government, Education, the Fine Arts, and Astronomy. Looking away from London towards the west are the statues of Agriculture holding a scythe, Architecture holding a model of St Paul's Cathedral, Engineering holding an engine and Pottery holding a vase.

The bridge is attractively coloured in burgundy and orange paint work. Adjacent to the bridge (in the direction of the Tate Gallery) stands a small riverside garden adorned by a sculpture of Henry Moore's called 'Looking Piece'.

Chelsea Bridge.

A full moon and floodlighting illuminate Chelsea Bridge and Battersea Power Station.

Low tide at Vauxhall Bridge.

LAMBETH BRIDGE

Ancient documents refer to the 'Great Bridge' at Lambeth, causing confusion as the exact date on which a bridge was built here. In fact the bridge referred to was a landing stage capable of receiving the monarch on state occasions. This landing stage was certainly in use as far back as the fourteenth century when Edward III levied taxes on wool and leather in order to finance the building of this structure. It was also mentioned as a meeting place between Henry VIII and Cranmer, and Queen Elizabeth I and Archbishop Parker.

For many years the Archbishop of Canterbury operated a ferry between Westminster and Lambeth. After Westminster Bridge was built in 1750, the Archbishop surrendered his lease on the ferry as, at that time, Westminster was sufficient for the needs of the local communities.

As the population of Lambeth grew, however, the people felt that another bridge should be built. They argued that they should not have to cross at Westminster, but should have a bridge closer to the locality.

An Act authorising a bridge was first passed in 1809, but insufficient money was raised so the Act lapsed. In 1860 the

Dusk at Lambeth Bridge, with the Palace of Westminster beyond.

Lambeth Bridge company finally obtained both a new bridging Act and sufficient funds to pay for a suspension bridge.

It was opened in 1862 and freed from tolls in May 1879, after it had been taken over by the Metropolitan Board of Works. By that time the ironwork of the bridge had rusted and become unsafe. It was decided to build a new bridge. Work began in June 1929. It was constructed of steel and reinforced concrete, with polished granite facings. King George V and Queen Mary opened the new structure in July 1932.

WESTMINSTER BRIDGE

For centuries London Bridge was the only stone bridge over the Thames in London itself. The next bridge up-river was at Kingston, many miles away.

The need for a bridge at Westminster had long been a source of dispute. It was possible to cross the river at this point by 'The Lambeth Horseferry'. This was inconvenient and could be unpleasant for several reasons. The ferry took a long time, could be dangerous when the tides ran swiftly and, additionally, the Lambeth horseferry was notorious for the bad language of the watermen! In 1701 an order imposed fines on the watermen because it said 'they do often use such immodest, obscene and lewd expressions . . . as are offensive to all sober persons, and tend extremely to the corrupting of Youths'. Despite the fines, the swearing continued.

Several attempts had been made to pass an Act authorising a bridge at Westminster in the time of Elizabeth I, James I and Charles I, but the City always drew up a long list of objections. They thought London Bridge was adequate; that the cost of building another bridge would make less money available for the maintenance of London Bridge; that the currents would run still faster; that a new bridge would encourage more buildings and sewage in Westminster. The list was endless.

Finally, in 1736, the Earl of Pembroke and his followers were granted their Act. It stated that 'a bridge would be advantageous not only to the City of Westminster but to many of His Majesty's Subjects and to the Publick in general'. The Act empowered certain Commissioners to raise money by holding a lottery. This scheme was violently opposed by the City and the watermen. They named it 'the Bridge of Fools' for they said that its foundations were built on a gamble.

Having agreed to a bridge, the next cause of dispute was the question of materials. The city wanted to have a timber bridge because it would be cheaper to build. Others, such as Nicholas Hawksmoor, suggested that a stone bridge would withstand the tides better than timber.

Eventually, the man chosen to build the bridge was Charles Labelye, a Swiss Engineer. He proposed using cast-iron caissons driven into the river bed by an engine invented by M. Valoue, a Swiss watchmaker. Instead of driving elm piles into the bed for foundations each caisson was to form the foundation. The bridge was to have fifteen semi-circular arches, giving an overall bridge length of 1,068 ft and 44 ft in width. The Earl of Pembroke laid the foundation stone for the first pier in January 1739.

Work went well through the spring and summer of that year, but the winter that followed brought the severest frosts that London had seen for many years. The Thames froze solid. The people of London walked on the river. They held a great Frost Fair at which whole oxen were roasted on the ice. It was the first of a series of hard winters, which delayed progress on the bridge. As

A classic view of Big Ben and Westminster Bridge at night – 10.55 p.m. to be precise.

the work dragged on, so the money to pay for it became scarce. The lotteries lost their popularity and the government was asked to contribute to the cost.

In addition, there had been many accidents during construction which also added to the cost of the bridge. When the foundations were being laid, barges often rammed into half-finished sunken piers which were not visible at high tide. This led to lawsuits and costly compensation to the barge owners.

Problems continued throughout the building of Westminster Bridge. Cracks appeared in the masonry. In 1747 stones from the fifth arch fell into the river. Two years later the Earl of Pembroke died. Having fought for the establishment of the bridge and given Labelye stout encouragement throughout the troubled years of building, he died just a short time before the bridge was opened in November 1750. Labelye himself was worn out by his worries and his hard labour throughout the ten years. His health was undermined and he retired to the warmer and kinder climate of the South of France, where he died in 1781.

Hungerford Railway Bridge.

Wordsworth wrote his famous poem to London from Westminster Bridge in 1802. The scene and the bridge itself were romantically observed. But the bridge was not a success. Labelye's method of laying piers without solid piles spelled disaster.

Eventually, in 1853, the decision was taken to build a new Westminster Bridge. Thomas Page and Charles Barry were appointed as engineer and architect respectively. Page ensured that the foundations would be strong and long lasting by driving massive elm piles into the river bed, onto which he added sheet iron and concrete.

The bridge is 827 ft long and 85 ft wide and was opened in May 1862. It has required very little repair since then.

HUNGERFORD RAILWAY BRIDGE

The site of Hungerford Bridge was first spanned by a suspension footbridge built by Brunel. Opened in 1845 the 14 ft wide and 1,440 ft long footbridge saw great use, particularly after Waterloo station opened in 1848. By 1860 the South Eastern Railway (SER) wanted to extend its line from London Bridge to the Strand, crossing the Thames on the same site as the suspension bridge.

At that time the SER resident engineer, Sir John Hawkshaw, was involved in the completion of the Clifton Suspension Bridge, in Bristol, in memory of its designer, Brunel, who died in the previous year. The engineer devised an ingenious scheme whereby the chains and other suspension elements of the Hungerford footbridge were removed for use on the Clifton Bridge, leaving the abutments and piers for the new railway crossing.

It was fully opened on 1 May 1864, together with a toll footbridge on the downstream side. Tolls, however, were abolished in October 1878 when the SER received £98,540 from the Metropolitan Board of Works for the maintenance of a permanent footway.

Despite having been widened in 1882 by Francis Brady (then the SER's engineer) and having its iron girders replaced by steel in 1980–81, two brick piers still remain as tangible reminders of Brunel's suspension footbridge.

Early morning at Hungerford Bridge.

WATERLOO BRIDGE

Although the clear, uncluttered lines of this famous bridge are world renowned, it is the views from either side of the bridge which are spectacular.

Whether by night or by day, the London skylines are instantly recognisable and beautiful. Looking westwards at night, the threads of light along the Embankment lead the eye to Big Ben and the Houses of Parliament. To the east one sees the illuminated dome of St Paul's Cathedral and the spires of other City churches, together with the more modern skyscrapers of the financial heart of London. On the South Bank stands the Royal Festival Hall and the National Theatre, separated by the bridge itself.

Lying in the shadow of the bridge, close to the northern bank there are the boats of the Thames Division of the Metropolitan Police by their floating police station. They patrol the river between Teddington and Dartford.

Completing the riverscape is the newest structure, the Festival Pier. Opened in 1983, this pier leads from the gardens facing the Festival Hall on the South Bank. As the first new pier built for many years it is a further indication of the river's tourist appeal – it is part of a network of piers that are served by passenger steamers from Greenwich upstream to Hampton Court and on up as far as Oxford.

The first Waterloo Bridge was built between the years 1811 and 1816, Its original name was the Strand Bridge, having been promoted by the Strand Bridge Company. The Act authorising its construction was passed in 1809 and John Rennie was chosen as designer. It was to span the Thames from close to Somerset House on the north bank to Waterman's Stairs on the south bank. The overall length of the bridge was 2,346 feet.

Rennie's design was of nine arches, constructed of granite and decorated by a pair of Doric columns on each pier. When the bridge was officially opened by the Prince Regent, it was re-named Waterloo Bridge, as the date – 18 June 1817 – was the second anniversary of the Duke of Wellington's famous battle.

In 1878 the bridge was bought by the Metropolitan Board of Works and was freed from tolls at a ceremony performed by the Prince and Princess of Wales. On inspection of their acquisition, the Water Board discovered that the foundations were in great need of repair. Each pier was duly reinforced between 1882 and 1884.

Waterloo Bridge.

Dawn silhouettes of Blackfriars Bridge, St Paul's Cathedral and the City of London.

By 1923 it had become evident that the repairs of forty years before were insufficient to save Rennie's bridge. Then followed a hot dispute! The Conservationists wanted the original bridge restored. London County Council considered this futile as the bridge (which was only 27 ft wide) would not be wide enough to carry the increased and heavier traffic. They wanted it replaced. The government, for its part, refused to make any financial commitment to the project.

A temporary structure was thrown across the river, as by 1923, the bridge was pronounced unsafe. But the arguments still continued. In 1934 the London County Council decided that the 'temporary' situation could no longer continue. They contracted Sir Giles Gilbert Scott, the architect, and the engineers Rendel, Palmer & Tritton to replace both the disused and temporary bridges with a new structure.

The foundation stone (cut from the old bridge) was laid in 1939. Although building the new bridge was given priority, its completion was delayed by the start of World War 2. Moreover, it was damaged by enemy action on many occasions – the only Thames crossing to suffer so. Despite this, its six lanes were opened in 1942.

BLACKFRIARS BRIDGE

The Black Friars were a Dominican order of monks, robed in black. They begged alms in order to survive the very barest of existences. In 1274 they moved their monastery from Holborn near to the site from which the road approaches the bridge. In their memory it is called Blackfriars, although that was not the original name for this bridge. When it was constructed, it was named after William Pitt. Indeed the tin plate buried with coins in the foundation stone is engraved with Pitt's name.

In 1753 the City Corporation decided that Blackfriars Bridge

Blackfriars Bridge and the City of London skyline from Waterloo Bridge.

should be built across the Thames to make a new gateway to the city. It was only the third bridge to be constructed in the vicinity, after London and Westminster. The Corporation invited competition entries for the best design for the new bridge. Robert Mylne, one of the sixty-nine entrants won the competition in 1759, and his appointment was confirmed in 1760. The Lord Mayor laid the foundation stone in June of that year. The nine arches were faced in Portland Stone and the bridge was opened in 1769.

Originally, tolls were charged with which to compensate the Watermen who had lost part of their livelihood each time a new bridge was built. Eventually the tolls were abolished in 1811.

Even regular maintenance of the bridge did not prevent the Portland stone being eroded by the water, and the bridge became seriously damaged by the scour of the polluted water. Part of the pollution was caused by the River Fleet which ran out close to the western end of the bridge under a large archway. In the years when the Fleet was large enough for barges to travel up as far as Holborn, it ran swiftly. But gradually, as it stagnated due to its constant use as a sewer, it poured its heavily polluted water into the Thames, thereby damaging the foundations of Blackfriars Bridge, as well as causing a health hazard.

A new bridge was designed and built by Joseph Cubitt. It has five arches, each supported by a cutwater with a semi-circular pulpit, decorated with sculpted birds and plants, as pedestrian refuges. The bridge is a solid structure of iron caissons filled with concrete. The brickwork is faced with granite and heavy wrought-iron ribs under each of the arches. It was opened by Queen Victoria in November 1869, on the same day as she opened Holborn Viaduct.

BLACKFRIARS RAILWAY BRIDGE

The London Chatham & Dover Railway was authorised in 1860 to build an extension from its existing terminus in Beckenham to Ludgate Hill in the City of London. The line was to cross the Thames close to Blackfriars Road Bridge and, as the latter was about to be rebuilt, it was decided that Joseph Cubitt should design both bridges.

Blackfriars road and rail bridges.

For the railway he designed a five-span wrought-iron bridge. Most notable are the ornate capitals and the large shields that incorporate the railway company coat of arms that were built on either end of the bridge. The bridge, together with Blackfriars station, was opened in December 1864.

This first bridge only took four tracks and when a new station was opened at St Paul's (later Holborn Viaduct) and the tunnels at Snow Hill were built to allow through working onto the Midland lines at Faringdon, it was decided to build a second bridge alongside the first. W. Mills, the resident engineer, designed a seven-track structure matching the spans of the earlier bridge. It was opened in May 1886.

Unfortunately, since then, it has been downhill all the way, for both the railway company and its services across the Thames. When the railways were grouped in 1923, the newly formed Southern Railway concentrated all the long-distance and Continental traffic at Waterloo and Victoria, thus stripping Blackfriars of all but local and suburban services. Its past glories are remembered, though, in the reconstructed station, where the original Portland Stone pillars announcing the exotic destinations have been restored and preserved. As for the two bridges, in 1971 all rail services were concentrated on the downstream span. The original structure, by then the oldest remaining railway bridge over the Thames in London, was eventually dismantled during 1984 – at a cost of £800,000.

SOUTHWARK BRIDGE

Roman London had three riverside harbours on the northern bank of the Thames. Barges loaded and discharged their goods within the City walls at Queenhithe, Dowgate and Billingsgate (the old fish market). All three have declined as the Port of London grew in importance. In fact, Queenhithe (hithe being an ancient word for harbour) is no more than a narrow alleyway between warehouses leading down to the Thames. It is 300 yards east of this ancient harbour that Southwark Bridge spans the Thames.

The Southwark Bridge Company was founded in 1813 because the people felt the need for another bridge. This was to no small extent due to the fact that London was fast expanding as a result of the Industrial Revolution and the growth of its population. John

Southwark Bridge.

Rennie was chosen to design the new bridge. It was the second of his London bridges and was made of three flat cast-iron arches. The City Corporation and the Thames Conservaters had objected to a bridge being built between Blackfriars and London Bridge. They claimed that it would be hazardous for navigation to have three bridges so close together.

However, John Rennie's design was such that the central arch was so wide as to give minimum obstruction to any river traffic passing beneath it. The new bridge was opened at midnight on 24 March 1819. As it was a toll bridge and both Blackfriars and London Bridge were free of tolls, Southwark was an unpopular and little used crossing. In 1864 the decision was taken to free Southwark from tolls so that it could share the increasing burden of traffic with the other two bridges.

Neither the approach roads nor the width of the bridge were sufficient to meet the needs of the traffic. Early in the twentieth century, it was decided to replace Southwark. Demolition began in 1913, but due to the start of World War 1 completion of the new bridge was delayed until 1921. It was designed by Basil Mott in conjunction with Sir Ernest George, the architect. Opened by George V on 16 June 1921, it has five steel arches supported by turreted granite piers. The bridge is lit by attractive lampstands, each bearing three lamps. It has recently been repainted in blue and white similar to Blackfriars Bridge.

CANNON STREET RAILWAY BRIDGE

The South Eastern Railway extended their system into the City of London less than one mile north from their original London Bridge terminus. Sir John Hawkshaw designed both Cannon Street station and the bridge over the river. The bridge has five spans, the navigable ones each being 167 ft long. When opened the bridge carried five tracks and two footpaths, the upstream path being open to the public on payment of a halfpenny toll.

In 1889 Francis Brady widened the bridge to accommodate more tracks and the public footpath was swept away. He doubled the original pairs of cast-iron cylinders to four on each pier. Since then the bridge has been rebuilt twice – shortly before World War

Recently modernised Cannon Street Railway Bridge.

2, and as recently as 1981, when concrete decking and cladding replaced some of the ironwork.

LONDON BRIDGE

Without London Bridge could London have developed into the city that it is today? – world centre for commerce, for the arts, seat of government, and home of 7 million inhabitants. To find out if this bridge has been such a major influence on the growth and importance of the city one must look back to early times and, initially, to its geographical position.

The Thames is a tidal river with safe inland waters, easily approachable from the sea. Ships could sail up the comparatively calm waters from the North Sea. Both on the northern and southern banks, at the site chosen for London Bridge, the land was firm and suitable for settlement, whereas most of the banks were marshy.

We know that the Romans must have built timber bridges on this site because in 1834 wooden piles set in iron bases were found. These were of the type that the Romans used. Also, Roman coins were dredged up nearby from the bed of the river.

There was no mention of a bridge in London until the time of Aethelwold, who died in AD 984. In the Codex Diplomaticus of those days a story is told of a woman who was accused of witchcraft and was drowned at London Bridge.

Even though the bridge would have been a light timber construction, it must have created an effective barrier against invaders such as the waves of Danes who repeatedly sailed up the Thames to fight the Anglo-Saxons. One Dane was not easily deterred. King Olav sailed right up to the bridge. His oarsmen were protected with a cover. Having reached the piles supporting the bridge, he destroyed them. The entire timber structure with its Anglo-Saxon defenders collapsed into the water. A poem which King Olav wrote is believed to be the origin of the ancient nursery rhyme – 'London Bridge is falling down'.

Later, in the time of King Canute, the bridge was again the main defence against invaders. Once more, in the fourteenth century and in 1677, when the British were at war with the French and Dutch respectively, the bridge proved to be a strong defence against the enemy.

Until 1209 the successive bridges were always constructed of timber. The expense of maintaining timber bridges was too costly. Money for this purpose was raised by taxing the people. During the twelfth century, they resolved to build a stone bridge. Peter, the Curate of Colechurch, was given the task of designing and constructing the new London Bridge. It was begun in 1176 and building continued for thirty-three years. The bridge had nineteen pointed arches, was 926 ft long and 40 ft wide. A wooden drawbridge (in the seventh span from the Southwark side) and piers rested on long platforms – called starlings – made of elm piles covered with thick planks.

On one of the piers a two-storied chapel was built. The Thomas à Becket Chapel was so well constructed that paper stored in the crypt below the two storeys was kept completely dry – at water level. Another interesting use was made of the starlings. Some were netted over and used as fish ponds. When the tide ran out fish were trapped under the netting and collected by the fishermen.

Verticals in perspective at London Bridge.

It was often said that London Bridge was built on woolpacks. This was because a tax on wool was levied towards payment of the bridge's construction. In 1212, three years after completion, the bridge was partly destroyed. A fire had broken out in Southwark – the wind carried it to the bridge – and 3,000 people perished caught in the flames, or drowned when jumping off the bridge. After this King John granted tolls for its repair.

Simon de Montfort marched on London in 1263 and to prevent his entrance to the city the bridge gates were locked and the keys thrown into the water. However, the crowd burst open the gates, and he entered London. In 1281 four of the arches were carried away by the flood.

In 1390 there was great pomp and ceremony on the bridge. A joust between the English Ambassador to Scotland and a Scottish Knight took place in the presence of Richard II and his court. The Englishman won!

The bridge was often a cause of excitement, not merely by events on its deck, but also under its arches. Because there were nineteen narrow arches, at high tide the water rose to such great heights that it gushed through the arches like waterfalls. This led to daredevils 'shooting the bridge', which ended with many a drowning. In 1428 the Duke of Norfolk's heavily laden barge struck a starling and sank – the duke and two of his gentlemen clung to the starling and were saved. Everyone else drowned. There was a saying that wise men walked over the bridge and only fools went under it. Certainly many people disembarked before the bridge, and reboarded after the bridge. The watermen themselves were not so fortunate and many drowned.

Southwark gate, guarding the southern end of the bridge, gained an unsavoury reputation. For many years it was decorated by spikes topped with heads of people executed at the Tower. In 1437 the great stone gate and two arches fell into the river, taking with them the gates' gruesome decorations. However, it was rebuilt including the spikes. Henry VII and Henry VIII kept the gate well ornamented, including the head of the scholarly Sir Thomas More. His daughter, Margaret, bribed a man to remove it, and she came by boat at night to take it away.

In Elizabeth I's time the bridge was restored and new buildings added to it. Over the years many houses had been built on the bridge, as the air was supposed to be healthier than in the city where the stench of sewage was often unbearable. One famous building was imported from Holland in 1577 in its entirety and rebuilt between the seventh and eighth arches. It was called Nonsuch House, and was made of carved and gilded timber, and stood four storeys high. Also during Elizabeth's reign a Dutchman, Peter Maurice, designed and constructed a water-mill at the northern end of the bridge.

Most of the houses on the bridge belonged to pin and needlemakers, printers and booksellers. In 1632, forty-two houses were burnt down due to the carelessness of a housemaid. She had put a tub containing hot ashes under some stairs at a Needlemakers. A few years later the Great Fire of 1666 razed several more houses.

By the middle of the eighteenth century the old bridge and its buildings were in a dilapidated state. The houses over the road almost blotted out the light on the bridge. Some of the structures had to be supported by wooden arches. Between 1750 and 1760 it was decided that both for health and safety reasons the buildings should be removed from the bridge. During demolition three pots of Elizabethan money were found in the old ruins.

In 1758 a temporary footbridge was thrown across the Thames while repairs were in progress, but it was destroyed by fire. Repairs were calculated at £2,500 annually – a great deal of money in those days. Furthermore, about fifty watermen were drowned in accidents each year. It was evident that a new bridge was essential.

In 1823 the City Corporation decided on a new bridge to be built 100 feet westwards from the old one. John Rennie was chosen to design the bridge. The first stone was laid in 1825 by the Lord Mayor, in an elaborate ceremony, in the presence of the Duke of York and 2,000 invited guests who were all seated in a specially constructed grandstand.

The bridge was opened six years later on 1 August 1831. King William and Queen Adelaide in their state barge, followed by a fleet of smaller vessels, travelled down river and landed on the northern steps of the bridge. The new bridge had only five arches which made navigation much safer than under the previous nineteen. At over 1,005 ft long and 56 ft wide it cost £1.5 million.

The old bridge, which had lasted so many years, left few mementoes. In the mud of the Thames lie three of the City gates, the ruins of the chapel and the bones of Peter of Colechurch. His bones were found when the chapel was demolished. Why were the bones not re-interred in some suitable place? He was, after all, one of the greatest ecclesiastical architects of his day.

Rennie's bridge lasted 140 years before the present bridge was built. The City, although loath to destroy the last of Rennies' city bridges, had to face the fact that the bridge was not solid enough, nor wide enough, to cope with the heavy traffic of the twentieth century.

The city engineer, H.W. King, together with Mott, Hay & Anderson and the architects William Holford & Partners designed a bridge which was to be made of pre-cast concrete and laid in four strips forming a roadway 105 ft wide. It has six traffic lanes and two pedestrian footpaths, one 15 ft wide and the other 20 ft wide. The parapets are of polished granite. The upstream and downstream face are floodlit below deck level. Road heating was laid within the surface, designed to operate automatically at low temperatures so as to prevent the icing of the roadway. In 1967

John Mowlem and Co. Ltd were awarded the contract to demolish the Rennie bridge and build the new London Bridge. The new bridge was opened by Queen Elizabeth II on 16 March 1973 in a much quieter ceremony than the previous bridge had witnessed for its opening.

Mr King was given the task of selling the old London Bridge and, surprisingly enough, he received five enquiries. The bridge was eventually dispatched to Lake Havasu in Arizona, USA, cut into sections to be re-constructed on arrival. It entered US customs labelled as 'large antique'. Small pieces were sold to individual buyers all over the world. Two slabs of the bridge reside at the edge of the ornamental lake in Kew Gardens. When the bridge was rebuilt in Arizona, the Lord Mayor of London, Sir Peter Malden Studd, was invited to perform the opening ceremony. Since then, London Bridge days are held annually in the first ten days of October. The people of Havasu dress in Elizabethan costumes and hold pancake races, archery contests and perform square dancing!

The new London Bridge has also had an eventful life to date. It is occasionally the scene of the traditional walking of a flock of sheep into the City – a right of all who are Freemen of the City of London. Even more bizarre, though, was the incident in June 1984 when a Royal Navy warship crashed into the downstream side of the bridge! HMS Jupiter had been residing in the Pool of London alongside HMS Belfast for a week. Facing upstream, the warship had to move away from Belfast and turn about in order to return through Tower Bridge to the North Sea. In this manoeuvre, the captain of HMS Jupiter misjudged the strength of the tide and hit London Bridge broadside. Damage to the ship was fairly extensive – particularly to the superstructure. The bridge did not escape damage either as the force of the ramming dislodged the granite parapet by eight inches!

TOWER BRIDGE

By the middle of the nineteenth century it became evident that a bridge was needed to serve the many people who lived east of London Bridge. Every day thousands of vehicles and pedestrians

Sunrise at Tower Bridge.

had to make lengthy detours to cross the river at London Bridge. In 1876 the City Corporation set up a special Bridge Committee to promote a new crossing.

The initial problem was how to erect a bridge that would allow the passage of tall ships to reach the Port of London. Sir Horace Jones, the City engineer, suggested a bascule bridge in 1878. Seven years later the City Corporation agreed to a modified design. Later that year (1885) the Act authorising the construction of Tower Bridge was passed. The Government of the day made one stipulation, that the bridge should match the style of the Tower of London, which dates back to 1078.

V.S. Pritchett wrote of Jones and the bridge he built: 'Asked to build an epoch-making suspension bridge that shall be a draw-bridge as well, he builds in the nineteenth century something like a medieval castle of granite ... and at the peak point of London's power and modernity, he creates a bridge suitable for King Arthur, the Black Prince, the archers at Agincourt ... Whatever one may say about the design of Tower Bridge, it is a London

Tower Bridge showing the roadway arch on the south side of the river.

landmark which is world renowned. How many thousands of tourists send home picture postcards of Tower Bridge, or take home a souvenir plate decorated with its picture? It is to London what the Eiffel Tower is to Paris – instantly recognisable.

Work on the bridge began in 1886 and it took eight years to construct. Sir Horace's original design was modified and the work was carried out by Sir John Wolfe Barry, assisted by Brunel (the younger) and Crutwell, the resident engineer. The central bascules of the bridge, each weighing about 1,000 tons were raised by an amazing feat of Victorian engineering. Having been installed at the bridge's inception, they were not replaced until 1976. In that year new electrically operated machinery was installed. As tall ships no longer go beyond Tower Bridge the roadway is seldom raised. Although in the first few months the bascules were raised 655 times, nowadays they are only raised three or four times each week. Advance notice is given to the Bridge Master before the bascules can be raised. More than 200,000 vehicles cross the roadway each day. Once, in 1952, a London bus was caught at an awkward angle between the opening halves of the roadway. It takes three to five minutes to open the bascules and road traffic is now regulated by traffic lights so that accidents like the one which befell the bus should not happen.

When Tower Bridge was built, the authorising Act stipulated that the public had to have access over the bridge at all times. To this end walkways were built 140 ft above the Thames between the two towers. In 1910 the walkways were closed because of the number of suicides that had taken place when people jumped off into the river. Additionally, it was felt that since opening the bascules had been reduced to six minutes, this would not constitute too long a delay for pedestrians.

In 1982, at a cost of £2.5 million, Tower Bridge was renovated and repaired. The walkways were re-opened as a tourist attraction. On the south side of the bridge there is a museum which displays a complete history of events connected with the bridge since its opening in 1894. The two original steam-driven pumps used for opening the bascules can be seen in the museum.

BIBLIOGRAPHY

Anderson, J.R.L. (1970), *The Upper Thames*, Methuen.
Biddle, Gordon & Nock, O.S. (1983), *The Railway Heritage of Britain*, Michael Joseph.
Burstall, Patricia (1981), *the Golden Age of the Thames*, David & Charles.
Chaplin, Peter H., *The Thames from Source to Tideway*, Whittet Books.
Christiansen, Rex (1981), *A Regional History of the Railways of Great Britain Volume 13 Thames & Severn, David & Severn*, David & Charles.
Cracknell, Basil E. (1968), *Portrait of London River*, Robert Hale Ltd.
Dunbar Janet (1966), *Prospect of Richmond*, Harrap.
Frowde, Henry (1910), *A History of Abingdon.*
Hall, Mr & Mrs S.C. (1859), *The Book of the Thames*, Charlotte James.
Harper, Charles (1910), *Thames Valley Villages*, Chapman & Hall Ltd.
Hollis, Charles, *Proposed Improvements in Lambeth and Westminster.*
Howard, Philip (1975), *London's River*, Hamish Hamilton.
Humphrey, Arthur L. *Caversham Bridge (1231–1926).*
Jebb, Miles, *Thames Valley Heritage Walk* (A Constable Guide).
Jeffrey, J. (1906), *Notes on Old Chelsea*, Home Counties Magazine.
Jenkins, Alan (1983), *The Book of the Thames*, Macmillan.
Mayo, Walter Langley Bourke (1929), *Report on The Thames from Cricklade to Staines.*
Mead, (1970), *London Bridge, Past, Present and Future*, London Society Journal.
Merson, George, *The illustrated Guide to the Great Western Railway*, (1852).
Morley, F.V. (1926), *River Thames*, Methuen.
Phillips, Geoffrey (1981), *Thames Crossings*, David & Charles.
Pritchard & Carpenter, M. and H. (1981), *A Thames Companion*, Oxford University Press.
Pudney, John (1972), *Crossing London's River.*
Ryan, E.K.W., *Thames from the Towpath.*
Senior, William (1891), *The Thames from Oxford to the Tower.*
Simmons, Jack (1973), *London Bridges of the Past*, London News.
Sweett, Cyril & Partners, *London Bridge Through the Ages.*
Taylor, Joseph (1914), *Great Britain States, 1763.*
Thacker, Fred S. (1920), *The Thames Highway Volumes I and II.*
Walker, J.W. (1931), *A History of Maidenhead.*
Walker, R.J.B., *Old Westminster Bridge.*
Ward Lock Red Guide, *Thames Valley and Oxford.*
Welch, Charles (1894), *History of Tower Bridge.*
Whishaw, Francis, *History and Construction of Westminster Bridge.*
White, H. (1971), *A Regional History of the Railways of Great Britain, Volume 3 Greater London*, David & Charles.
Wilson, David Gordon (1977), *The Making of The Middle Thames*, Spurbooks Ltd.

INDEX

Numbers in **bold** indicate the main text entries and bridge illustrations.